STATISTICAL ANALYSIS
WITH ARCVIEW GIS®

STATISTICAL ANALYSIS WITH ARCVIEW GIS®

JAY LEE
DAVID W. S. WONG

JOHN WILEY & SONS, INC.

New York / Chichester / Weinheim / Brisbane / Toronto / Singapore

Library of Congress Cataloging-in-Publication Data

Lee, Jay.
    Statistical analysis with ArcView GIS / Jay Lee and David W. S. Wong.
      p.   cm.
    Includes bibliographical references (p.).
    ISBN 0-471-34874-0 (cloth: alk. paper)
    1. Geographic information systems.   2. Statistics.   3. ArcView.   I. Title: Statistical analysis with ArcView GIS.   II. Wong, David W. S.
    III. Title.
    G70.212.L43   2000
    $910'.285$—dc21   00-032090

Printed in the United States of America

10 9 8 7 6 5 4 3 2 1

CONTENTS

INTRODUCTION

The proliferation of *geographic information science and systems (GIS)* has prompted researchers in many fields to reconsider their ways of conducting research or solving practical problems. GIS tools have enabled users to effectively capture, manage, analyze, and display their geographically referenced data.

With GIS, researchers can now process larger volumes of data within a shorter time and with greater precision. Furthermore, similar analytical processes can be easily repeated for different data sets. What used to be labor-intensive tasks are now often performed within minutes or even seconds with computers. Researchers are less likely to be limited by computation time. They do not need to avoid working with large volumes of data when they explore new ways of processing data. As a result, new approaches to exploring spatial and nonspatial data have been developed to execute processes that once were only dreamed of.

The commercialization of GIS technology also elevated the uses of GIS to a higher level than ever before. With packaged computer programs available to everyone at a reasonable cost, mapping complex geographic data and overlaying thematic data layers for site selections have become routine tasks that can be performed even by novice GIS users.

One important characteristic that separates GIS from other information systems is their ability to handle spatial data. Spatial data are sensitive to *scales* and to the way they are measured (scales of measurement). A city is only a point on a world map. The same city, however, occupies an entire sheet of map when all of its streets, expressways, rivers, and so on are displayed in detail. A river shown as merely a thin line on a continental map may become an areal phenomenon when it is mapped at a local scale. The ability to manage geographic objects across different scales has made GIS valuable tools for many research fields and applications.

Another characteristic that makes spatial data unique is that location information is embedded in the observations. In addition to the attributes describing the characteristics of the observations (geographic features), locations of features can be extracted from the data and analyzed. GIS are valuable tools to accomplish this.

In this book, we use a model of three types of geographic objects to simplify and symbolize the complex real world. These three objects are points, lines, and polygons. They represent different types of geographic features and phenomena. In addition, they are associated with attribute information to give meaning to the objects and to distinguish objects from one another.

The attribute information that describes the characteristics of geographic objects can be measured at different scales. Data measured as categories, for example, need to be treated differently from those measured as ordered sequences or values of ratios. Consequently, we would display them using different methods and symbols.

But GIS are not limited to the display of spatial data. They are most useful when used to perform data analysis. In shifting from spatial data management systems to spatial decision support systems, GIS have provided tools for spatial queries, dynamic mapping, geo-coding, and even simulating alternative scenarios of future regional development. Unfortunately, most GIS users use GIS only for mapping purposes or, at most, for buffering or overlaying various thematic data layers—rudimentary spatial analytical procedures supported by almost all GIS.

Given the need to draw inferences from empirical data, there have been increasing demands for combining *statistical analysis* with GIS so that geographic data can be processed, analyzed, and mapped within the same working environment. All GIS now offer functions to calculate simple descriptive statistics. However, users are calling for *spatial analysis* so that they can use spatial statistics developed specifically for spatial data.

Since spatial analysis is not widely available in the format of computer programs that work directly with GIS, problems or applications that should have been dealt with by spatial statistical analysis are being treated with tools from classical statistics. This is because these classical statistical tools are readily available from packages such as SAS^{TM}, SPSS^{TM}, and MiniTab^{TM}. Data sets that should have been explored at various geographical scales are being examined only at the scale of convenience because it is often very difficult, or sometimes impossible, to do so with conventional statistical packages. As such, classical statistical methods are being applied inappropriately in many cases, ignoring any spatial associations or spatial relationships existing among the data being analyzed. These relationships potentially violate assumptions essential to drawing inferences in classical statistics. As a result, when classical statistical methods are applied to spatial data, results may be biased.

Some attempts have been made to integrate spatial statistical analysis with GIS, but many of these attempts failed to encourage general, nonacademic users to apply them to real-world problems (e.g., Anselin, 1992; Anselin and Bao, 1997; Bao and Martin, 1997; Griffith, 1989, 1993). This is because these implementations were only loosely integrated with GIS packages or because they required additional transformation of data between different formats. When only loosely coupled with GIS, implementations of spatial statistical analysis are often too difficult to use because users must learn additional software packages to perform even a simple analysis. When they require transformation of data between different formats, implementations of spatial statistical analysis are often abandoned because most users cannot comprehend the tedious processes. Also, users no longer can work within an integrated environment to handle spatial data conveniently. For instance, if exploratory analysis of spatial data is desired, data in the GIS environment have to be transformed and moved. Analytical results of external procedures supported by popular statistical packages have to be brought back to GIS for visual display.

In this book, we provide our implementation of integrating spatial statistical analysis with GIS. We offer this implementation because we believe that GIS have matured to the point where spatial statistics can be brought into GIS by developing modular codes. Our implementation of spatial statistical analysis is completely integrated into the GIS package. We believe it will make spatial analytical tools and spatial statistics more accessible to users in various fields of research and applications.

We choose to implement spatial statistics in ArcView GIS^{TM} (ESRI, Redlands, California). This selection is based on years of teaching at each of our two universities and the significant market share ArcView GIS has compared to other alternatives. The selection is also based on the fact that ArcView GIS is frequently referenced in various research papers, journal articles, and conference presentations. However, while our implementation uses ArcView GIS, we believe that similar implementations can be carried out in other GIS packages.

With this book, we wish to disseminate our experience in applying spatial statistical analysis to various practical problems. For each method, the discussion of various aspects of the spatial statistics is followed by examples that show the steps of calculations, as well as ways to interpret the results. At the end of each section, we offer ArcView Notes to guide readers in using the statistics in ArcView GIS using project files we developed for each chapter. All of the statistical procedures we have added are written in ArcView GIS^{TM} Avenue scripts. These scripts are embedded in each project file and are accessed through the corresponding project files. Unless otherwise specified, these scripts in the project files use the first theme in the View document added by the user to perform the analysis. These project files are available in the accompanying website. Certain data sets used in the examples are also included.

Our implementation of spatial statistical analysis should enable users to perform spatial analysis with tools that are not yet available in popular GIS packages. We encourage readers to use them in a wide spectrum of application areas and use them correctly.

While this book was not developed as a textbook, it is possible for it to be used in a classroom setting or in a week-long workshop. We hope that the book is well organized enough so that readers can use it in self-paced study. When it is used in teaching courses, instructors may want to develop additional examples or exercises, related to their research areas, to complement the discussion in this book.

We have made no assumption about the statistical background of the readers of this book or the users of the accompanying programs. At the same time, we have avoided the inclusion of advanced statistical concepts associated with various topics being discussed in the book. We have tried to develope this book as a practical tool to bridge the gap between books on theories and software manuals. On the other hand, we assume that readers are familiar with the basic operations in ArcView, such as opening an existing project file, adding themes to a view, accessing the feature attribute table, and making maps via the Legend Editor.

In additional to simple descriptive classical statistics, the analytical tools covered in this book fall into several categories: centrographic measures or descriptive geostatistics for describing point distributions based upon bivariate statistics; point pattern analyses; directional statistics for linear features; network analysis; and spatial autocorrelation analysis. Most of these tools are descriptive and exploratory. We realize that the next logical step is to model spatial processes using the spatial regression framework, including the models discussed in Anselin (1988) and Griffith (1988). Unfortunately, most native programming environments in GIS, including that of ArcView, are not mature enough to fully incorporate those advanced spatial models. We also prefer not to include advanced geostatistics and various spatial surface modeling methods such as those covered by Bailey and Gatrell (1995). The reason is that this area of spatial modeling, including a family of kriging methods, is more effective and appropriate for a raster GIS environment (Burrough and McDonnel, 1998), and the framework adopted in this book is more convenient for a vector GIS environment. In addition, many of the geostatistical procedures have already been implemented in many raster-based GIS and have been used extensively by many geoscientists.

In this book, we will first discuss tools in GIS that can be used to analyze attribute data. This is followed by chapters that examine tools for analyzing points, lines, and polygons. The spatial analytical tools built into ArcView GIS will be demonstrated with examples whenever possible.

REFERENCES

Anselin, L. (1988). *Spatial Econometrics: Methods and Models*. Boston, MA: Kluwer Academic Publishers.

Anselin, L. (1992). *SPACESTAT TUTORIAL: A Workbook for Using SpaceStat in the Analysis of Spatial Data*. Technical Software series S-92-1. Santa Barbara, CA: NCGIA.

Anselin, L., and Bao, S. (1997). Exploratory spatial data analysis linking SpaceStat and ArcView. In M. Fischer and A. Getis (eds.), *Recent Development in Spatial Analysis*. New York: Springer-Verlag.

Bailey, T. C., and Gatrell, A. C. (1995). *Interactive Spatial Data Analysis*. Harlow Essex, Englang: Longman.

Bao, S., and Martin, D. (1997). Integrating S-PLUS with ArcView in spatial data analysis: An introduction to the S+ ArcView link. A paper presented at 1997 ESRI's User Conference, San Diego, CA.

Burrough, P. A., and McDonnell, R. A. (1998). *Principles of Geographical Information Systems*. Oxford: Oxford University Press.

Griffith, D. A. (1988). *Advanced Spatial Statistics*. Boston, MA: Kluwer Academic Publishers.

Griffith, D. A. (1989). *Spatial Regression Analysis on the PC: Spatial Statistics Using MINITAB*. Ann Arbor, MI: Institute of Mathematical Geography.

Griffith, D. A. (1993). *Spatial Regression Analysis on the PC: Spatial Statistics Using SAS*. Washington, DC: Association of American Geographers.

Support Website Information

The ArcView GIS Avenue scripts referenced in this text are available for download from `http://www.wiley.com/lee`. Two packages are available: a .zip file for PCs and *nix machines, and a .sit file for Macs. Simply download the desired archive, extract the files to your local drive, and navigate the folders as described in the ArcView Notes found throughout the text.

CHAPTER 1

ATTRIBUTE DESCRIPTORS

The world we live in is a complex and dynamic one. To better understand the world, we often need to reduce it to some simple representative *models*. We often construct models of the real world to decrease the complexity of any problems we study to a manageable level so that we can solve them. We use models of the real world to provide a static version of nature so that we can focus better on the issues at hand.

Geographers typically model the world with objects located at different places on the surface of the world. We use different types of objects to represent the complicated world. We formulate relationships between objects to simulate the dynamics of the world systems. Specifically, we use models to represent the world via *simplification*.

A map is an example of how the real world is modeled. As the map in Figure 1.1 shows, objects in the real world are represented by different *symbols*: *lines* show how rivers run their courses and how roads are connected, while *points* (small, solid circles and squares) and *polygons* (or rectangles of various sizes) show the locations of special interest.

In Figure 1.1, the point representing the county fairground is easily recognized because a text label accompanies it. Similarly, Cuyahoga River, State Route 14, and other highways are identifiable because each of these has a label to identify it. For various buildings represented by squares, however, there is no additional information to help map readers separate one from another to show what they are or what they are for.

We need additional information to give *meaning* to the symbols we use to represent the real world. Like the squares in Figure 1.1, symbols remain only symbols unless we associate them with additional *attribute* information. Lines are

1

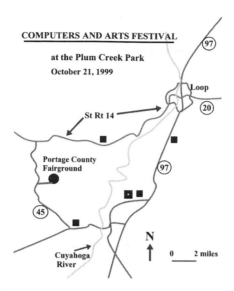

Figure 1.1 A map as a model of the real world.

only lines and points are only points if there is no additional attribute information to describe their properties and characteristics.

In managing geographic information, the conventional approach to structuring spatial data is to have *cartographic data* describing the locations, shapes, or other spatial characteristics of the objects and to have *attribute data* describing other characteristics of the objects. In Figure 1.2, a set of points, representing cities in the three-county area in northeastern Ohio, are shown. To describe each of these points, an *attribute table* records information on their characteristics. In this attribute table, each record is linked to a point. Each record contains a number of *fields* that store attribute data for the associated point. This way, the characteristics of each symbol we use in a map that represents geographic objects in the world can be described in detail in the attribute table.

ArcView The data structure described in this section is commonly known
Notes as *relational data structure.* Typically, a GIS database contains
 layers of thematic data on the area of interest. Each layer, represented as a shapefile in ArcView, has a map view and an attribute table. The map view is the cartographic data of the thematic layer where coordinates of points, lines, and polygons are used in displays. The attribute table, on the other hand, stores additional attribute information describing various aspects of the objects in the map view.

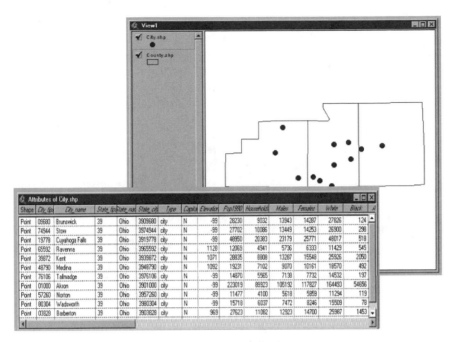

Figure 1.2   Map view and attribute table.

Let's first turn our attention to the attribute tables in geographic information systems (GIS) databases. As mentioned earlier, each record in an attribute table contains a number of fields. Since each record is associated with an object—or, in statistical terms, an observation in the map view—the data stored in the fields in a record are the information describing the associated object or observation in as many ways as the number of fields.

There are several types of data an attribute table can store. *Numerical data* are measured quantitatively. We can easily find examples of this type of data: areas of farms, precipitation amount at each monitoring station, population count for each county or city, and so on. This type of information is normally referred to as being measured at *ratio scale*. Data measured at ratio scale typically have a real zero value. For example, temperature at 0° Kelvin means no energy. By definition, there is no situation with a temperature below 0° Kelvin. Therefore, temperature measured in Kelvin is of ratio scale. Another obvious example is any data value measured in proportion, such as population densities, male/female ratios in high school classrooms, cost/benefit ratios, and so on. The absolute minimum for a proportion is 0. A proportion below 0 is not interpretable. Mathematical operations such as addition ($+$), subtraction ($-$), multiplication ($*$), and division ($/$) can be applied to data that are measured at ratio scale.

When there is no real zero value for the phenomenon being measured but the data are on a continuous scale, the data are measured at *interval scale*. Examples of this type of measurements include temperatures and elevations. A temperature of 0°C does not mean that there is no temperature. It simply means that the temperature is at the position where it was defined as 0°C. In fact, it is 32°F when the temperature is 0°C. For elevation, 0 meter above sea level does not mean that there is no elevation. It simply means that it is at the same elevation as the average elevation of oceans. For data measured at this scale, all four mathematical operations are applicable. With both ratio and interval data, putting aside the preciseness of the measurement, we know the exact position of the observation along the continuous value line. While interval data are measured by some defined intervals, the differences between intervals sometimes are not proportional. For example, the difference between 80°F and 90°F is not the same as the difference between 80°F and 70°F in terms of how warm you feel, even though the difference is 10°F in both cases.

There are situations in which data simply give the order of the measured phenomenon. In this case, the data are said to be measured at *ordinal scale*. We can use 1, 2, 3, . . . , to represent the order or the ranking of cities in a state according to their population sizes. We can use descriptions or terms such as *high*, *medium*, or *low altitude* to represent the heights of mountains in a rough sense. Then observations are grouped into classes, and the classes follow an order. With ordinal data, mathematical operations such as +, −, ∗, or / cannot be applied. With only their ranks, we know which city is larger than another given city, but we don't know by how much. For ordinal data, we know the order of measured phenomena but we cannot add two measures to get another measure, that is, 1st + 2nd = 3rd.

Finally, we can measure phenomena in categorical form. This is known as measuring at the *nominal scale*. For this scale, no mathematical operations can be applied because nominal data only identify individual objects being measured. We don't even have the order between these objects, which we would know if they were measured at ordinal scale. We can easily think of many examples of data at this scale: street numbers of the houses along a street, telephone numbers of friends, flight numbers, zoning codes for different types of land use, and so on. Please note that the numbers at nominal scale, simply represent different things. They cannot be added or multiplied. Adding two telephone numbers will not result in another telephone number. Dividing house numbers by another house number is meaningless.

ArcView Notes   In working with an ArcView Table document, users need to understand the attribute data they use. For convenience, categorical data are often coded with numbers, such as 1, 2, 3, . . . for different types of land use. In those cases, the numbers should be treated as characters. FIPS and ZIP codes also consist of numbers, but these numbers should not be used in any numerical calculations. In ArcView, data measured at ratio or interval scales are of the type `number`, while data measured at ordinal

or nominal scales are of the type `string`. There are ArcView
functions such as `AsString` or `AsNumber` that can be used to
convert attribute data between numerical and string forms.

GIS data sets are often large. A thematic GIS data layer of land use can easily contain more than 2,000 polygons for a large area. In a study matching potential customers and a newspaper's distribution system, the associated attribute table can easily have over 10,000 records in a moderately sized metropolitan region. Therefore, understanding the data will not be simple. For meaningful analysis of attribute data associated with map views, statistics are needed to describe, to summarize, or to find the relationships between attributes and geographical features.

In the subsequent sections, we will first look at ways to calculate descriptive statistics of attribute data using ArcView. These descriptive statistics indicate various statistical properties, such as central tendency and dispersion of the data. Statistics depicting the relationship between attributes will also be discussed.

1.1 CENTRAL TENDENCY

Often the first step in analyzing a set of numerical data is to measure their *central tendency*. The concept of central tendency uses a representative value to summarize the set of numerical data. For example, an *average* family income from a census tract gives an idea of the economic status of families in that census tract. Using the average family income to represent all income figures in that census tract allows us to quickly get an overall impression of its economic status.

In surveying students at Kent State University about their means of commuting, we found that most of them drive to the university. That specific type of commuting choice as nominal data, therefore, is the *mode* of commuting of those students. When comparing housing prices between neighborhoods, we often use the housing price in each neighborhood that stands close to the middle of the range of prices. Comparing the *middle* prices between neighborhoods allows us to avoid the pitfall of comparing the highest housing price in one neighborhood to the lowest housing price in another neighborhood.

The concept of central tendency is applied in everyday life. People use average Scholastic Aptitude Test (SAT) scores of freshman classes to compare how well a college does between years or to compare how colleges stand in relation to other colleges. We use phrases such as *typical weather* or *typical traffic pattern* to describe phenomena that happen most often. These are just a few of the many examples we can find everywhere.

1.1.1 Mode

The *mode* is the simplest measure of central tendency. It is the value that occurs most frequently in a set of data; consequently, that specific value is also known as the *modal value*. For categorical or nominal data, the category that has the most observations or highest frequency is regarded as the mode. When working with ordinal data, the mode is usually the rank shared by two or more observations.

The modal value for interval or ratio data may not be very useful because that value may not occur more than once in a data set. Alternatively, researchers often *degrade* or *simplify* interval or ratio data to nominal scale by assigning individual data to one of the categories that they set up based on ranges of data values.

In Table 1.1, a total of 51 countries and territories in North and South America are listed by their population counts in 1990, areas in square miles, population densities, and categories of low/medium/high population density. The locations of the listed countries are shown in Figure 1.3.

To illustrate the use of mode and the effect of degrading interval/ratio data to nominal data, Table 1.1 first calculates the population density of each country by dividing the population count by its area. When examining the derived population densities, we cannot find a mode because no two or more countries have the same

TABLE 1.1   Population Density of Countries in the Americas

Country	Population	Area in Sq Miles	Population Density	Category
Anguilla	9,208	33	276	Low
Antiuga and Barbuda	65,212	179	365	Medium
Argentina	33,796,870	1,073,749	31	Low
Aruba	67,074	71	950	High
Bahamas, The	272,209	4,968	55	Low
Barbados	260,627	170	1,534	High
Belize	207,586	8,562	24	Low
Bermuda	59,973	15	3,941	High
Bolivia	7,648,315	420,985	18	Low
Brazil	151,525,400	3,284,602	46	Low
British Virgin Islands	18,194	63	290	Low
Canada	28,402,320	3,824,205	7	Low
Cayman Islands	31,777	107	297	Low
Chile	13,772,710	286,601	48	Low
Columbia	34,414,590	440,912	78	Low
Costa Rica	3,319,438	19,926	167	Low
Cuba	11,102,280	42,642	260	Low
Dominica	70,671	283	250	Low
Dominican Republic	759,957	18,705	415	Medium
Ecuador	10,541,820	99,201	106	Low
El Salvador	5,752,470	7,991	720	High
Falkland Islands	2,136	4,446	0	Low
French Polynesia	217,000	1,167	186	Low
Grenada	95,608	142	675	High
Guadeloupe	410,638	673	610	High
Guatemala	10,321,270	42,279	244	Low

TABLE 1.1   *Continued*

Country	Population	Area in Sq Miles	Population Density	Category
Guyana	754,931	81,560	9	Low
Haiti	7,044,890	10,485	672	High
Honduras	5,367,067	43,572	123	Low
Jamaica	2,407,607	4,264	565	Medium
Martinique	374,574	425	881	High
Mexico	92,380,850	757,891	122	Low
Montserrat	12,771	41	314	Medium
Netherlands Antilles	191,572	311	617	High
Nicaragua	4,275,103	49,825	86	Low
Panama	2,562,045	28,841	89	Low
Paraguay	4,773,464	154,475	31	Low
Peru	24,496,400	500,738	49	Low
Pitcairn Islands	56	21	3	Low
Puerto Rico	3,647,931	3,499	1,043	High
St. Kitts and Nevis	42,908	106	404	Medium
St. Lucia	141,743	234	606	High
St. Pierre and Miquelon	6,809	94	72	Low
St. Vincent and the Grenadines	110,459	150	734	High
Suriname	428,026	56,177	8	Low
Trinidad and Tobago	1,292,000	1,989	650	High
Turks and Caicos Islands	14,512	189	77	Low
United States	258,833,000	3,648,923	71	Low
Uruguay	3,084,641	68,780	45	Low
Venezuela	19,857,850	353,884	56	Low
Virgin Islands	101,614	135	755	High

population density value. If we really want to identify the mode for this data set, the data have to be degraded from ratio scale to nominal scale.

If we define population densities below 300 persons per square mile as low, those between 300 and 600 persons per square mile as medium, and those over 600 persons per square mile as high, we can see from the last column in Table 1.1 that low density is the mode of this set of population densities. With the mode, we now have an overall impression of the levels of population density in these countries.

1.1.2 Median

The *median* is another measure of central tendency. In a set of data, the median is the *middle* value when all values in the data set are arranged in ascending or descending order.

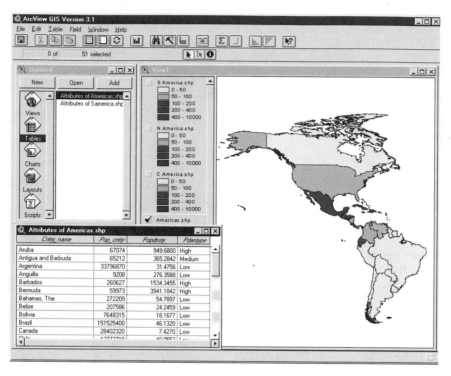

Figure 1.3   Population densities of the Americas.

To find the median of the population densities listed in Table 1.2, we first sort the table by population densities. Since 31 countries are listed in this table, the 16th value in the sorted sequence will be our median. The 16th entry in the list is 314 persons/mile^2 (Montserrat).

When the number of observations in a data set is odd, it is relatively simple to work out the median of the set. For a set of data containing an even number of values, the median is simply the value midway between the two middle values. For example, there are 12 countries listed in Table 1.3. The middle two values are 45 persons/mile^2 (Uruguay) and 46 persons/mile^2 (Brazil). The median of the set of 12 population densities will therefore be 45.5 persons/mile^2 since $(45 + 46)/2 = 45.5$ (persons/mile^2).

In general, a median can be found in any data set containing interval or ratio data. The median of a data set gives a value that is at the middle of the set. This median value is not severely affected by the inclusion of extremely large or extremely small values in the data set since it is defined by its position in the ordered sequence of data values.

1.1.3   Mean

The *mean* is the most commonly used measure of central tendency. It is the *average* value in a data set. This average is also known as *arithmetic mean* because of

TABLE 1.2   Population Density of Countries in Central America

Country	Population	Area in Sq Miles	Population Density	Category
Belize	207,586	8,562	24	Low
Bahamas, The	272,209	4,968	55	Low
United States	258,833,000	3,648,923	71	Low
Turks and Caicos Islands	14,512	189	77	Low
Nicaragua	4,275,103	49,825	86	Low
Panama	2,562,045	28,841	89	Low
Mexico	92,380,850	757,891	122	Low
Honduras	5,367,067	43,572	123	Low
Costa Rica	3,319,438	19,926	167	Low
Guatemala	10,321,270	42,279	244	Low
Dominica	70,671	283	250	Low
Cuba	11,102,280	42,642	260	Low
Anguilla	9,208	33	276	Low
British Virgin Islands	18,194	63	290	Low
Cayman Islands	31,777	107	297	Low
Montserrat	12,771	41	314	Medium
Antigua and Barbuda	65,212	179	365	Medium
St. Kitts and Nevis	42,908	106	404	Medium
Dominican Republic	7,759,957	18,705	415	Medium
Jamaica	2,407,607	4,264	565	Medium
St. Lucia	141,743	234	606	High
Guadeloupe	410,638	673	610	High
Netherlands Anitlles	191,572	311	617	High
Haiti	7,044,890	10,485	672	High
Grenada	95,608	142	675	High
El Salvador	5,752,470	7,991	720	High
St. Vincent and the Grenadines	110,459	150	734	High
Martinique	374,574	425	881	High
Aruba	67,074	71	950	High
Puerto Rico	3,647,931	3,499	1,043	High
Barbados	260,627	170	1,524	High

the way it is calculated. The mean is calculated by adding together all the values in a data set and then dividing the sum by the number of values. The equation for calculating the mean is

$$\overline{X} = \frac{\sum_{i=1}^{n} x_i}{n},$$

where $\overline{X}$ (read as "X bar") denotes the mean of a group of values: $x_1, x_2, \ldots, x_n$. If there were 5 values in the data set, n would be 5. The symbol, $\sum_{i=1}^{n} x_i$, means

TABLE 1.3   Population Density of Countries in South America

Country	Population	Area in Sq Miles	Population Density	Category
Argentina	33,796,870	1,073,749	31	Low
Bolivia	7,648,315	420,985	18	Low
Brazil	151,525,400	3,284,602	46	Low
Chile	13,772,710	286,601	48	Low
Columbia	34,414,590	440,912	78	Low
Ecuador	10,541,820	99,201	106	Low
Guyana	754,931	81,560	9	Low
Suriname	428,026	56,177	8	Low
Paraguay	4,773,464	154,475	31	Low
Peru	24,496,400	500,738	49	Low
Uruguay	3,084,641	68,780	45	Low
Venezuela	19,857,850	353,884	56	Low

adding all 5 values as follows:

$$\sum_{i=1}^{n} x_i = x_1 + x_2 + x_3 + x_4 + x_5.$$

As an example, even though it is simple, Table 1.4 lists the levels of population density for Canada and the United States. The mean can be calculated as

$$\overline{X} = \frac{\sum_{i=1}^{2} x_i}{2} = \frac{7 + 71}{2} = \frac{78}{2} = 39 \text{ (persons/mile}^2).$$

There are two density values, so $n = 2$. The mean is simply the average of the two values.

In Table 1.1, 51 countries are listed, so the mean population density is

$$\overline{X} = \frac{\sum_{i=1}^{51} x_i}{51} = \frac{x_1 + x_2 + \cdots + x_{51}}{51} = \frac{276 + 365 + \cdots + 755}{51} = 385.79.$$

TABLE 1.4   Population Density of Canada and the United States

Country	Population	Area in Sq Miles	Population Density	Category
Canada	28,402,320	3,824,205	7	Low
United States	258,833,000	3,648,923	71	Low

Similarly, in Table 1.2, the mean population density for Central American countries is

$$\overline{X} = \frac{\sum_{i=1}^{31} x_i}{31} = \frac{x_1 + x_2 + \cdots + x_{31}}{31} = \frac{24 + 55 + \cdots + 1534}{31} = 446.56.$$

For the South American countries, the mean population density can be calculated from Table 1.3 as

$$\overline{X} = \frac{\sum_{i=1}^{12} x_i}{51} = \frac{x_1 + x_2 + \cdots + x_{12}}{12} = \frac{8 + 9 + \cdots + 106}{12} = 43.82.$$

The above calculations of the mean of interval or ratio data are appropriate if all values are counted individually. But if observations are grouped into classes and all observations within each group are represented by a value, the calculation of the mean will be slightly different. The mean derived from the grouped data is usually called the *grouped mean* or *weighted mean*. Assuming that the value midway between the upper bound and the lower bound of each class is the representative value, x_i, and f_i represents the number of observations in the ith class, the weighted mean, $\overline{X}_w$, can be calculated as

$$\overline{X}_w = \frac{\sum_{i=1}^{k} f_i x_i}{\sum_{i=1}^{k} f_i},$$

where k is the number of classes.

Before computers were widely available, the grouped mean was used to estimate the overall mean in a very large data set. In this procedure, observations are divided into groups according to their values. A value from each group, typically the midpoint between the lower and upper bounds of the group, is used to represent the group. When calculating the grouped mean, the number of observations in each group is used as the weight. This is also the reason why the grouped mean is often called the weighted mean.

Compared to the median, the mean is very sensitive to the inclusion of extreme values. Even if only one extremely large value is added to the data set, the average of all values in the data set will be pulled toward a larger value. As a result, the mean may be overestimated.

It is important to note that mode, median, and mean are three different measure of central tendency. When applied to a common data set, these three measures will give three different values. They differ in their definitions and in how they are calculated, so they have different meanings.

1.2 DISPERSION AND DISTRIBUTION

While the mean is a good measure of the central tendency of a set of data values, it does not provide enough information to describe how the values in a data set are distributed. With the central tendency, we know what the average value is but we do not know how the values scatter around this average. Are the values similar to the mean, with only small differences? Do the values vary very differently from the mean? We don't know for sure unless we can measure how these values disperse or concentrate around the mean.

To illustrate the need for more information than the mean can give, let us use an example of the following series of numbers to compute their mean:

$$x_a:\quad 2, 5, 1, 4, 7, 3, 6$$
$$\overline{X}_a = \frac{2 + 5 + 1 + 4 + 7 + 3 + 6}{7} = 4.$$

The mean, 4, seems to be reasonably representative of these numbers. However, the following series of numbers also yields a mean of 4, with quite a different composition:

$$x_b:\quad 24, -18, 21, -43, 2, 33, -23$$
$$\overline{X}_b = \frac{24 + (-18) + 21 + (-43) + 2 + 33 + (-23)}{7} = 4.$$

If we only know the means of these two sets of numbers, and have no further information, we might speculate that the two data sets are very similar to each other because their means are identical. However, by briefly examining the two number series, we know that the first series has a relatively narrow range centered at the mean, while the second series has a very wide range, that is, a highly dispersed set of values. Relying on the mean alone to compare these two series of values will yield misleading results. The truth is concealed by the large positive and negative values offsetting each other in the second series.

To better understand how values in a data set distribute, a number of descriptive statistics can be used. These include mean deviations, standard deviations, skewness, and kurtosis. These measures provide information about the degree of dispersion among the values and the direction in which the values cluster. Together they describe the distribution of numeric values in a data set so that analysts can understand the distribution or compare it with other distributions.

1.2.1 Mean Deviation

The first measure of dispersion is the *mean deviation*. It takes into account every value in the data set by calculating and summing the deviation of each value from the mean, that is, the difference between each value and the mean. The equation for calculating the mean deviation is

$$\text{Mean deviation} = \frac{\sum_{i=1}^{n} |x_i - \overline{X}|}{n}.$$

For data series x_a, the mean deviation is

Mean deviation$_a$

$$= \frac{|2-4|+|5-4|+|1-4|+|4-4|+|7-4|+|3-4|+|6-4|}{7}$$

$$= \frac{2+1+3+0+3+1+2}{7} = 1.71.$$

The symbol $|x - \overline{X}|$ denotes the absolute difference between each value of x and the mean. So the equation first adds up all the absolute differences and then divides this number by the number of values to get the *average* of all absolute differences. This average absolute difference is the mean deviation. For the other series, x_b, the mean deviation is 20.57, which is quite different from 1.71 of series x_a.

This measure is simple to calculate and easy to understand. It provides a convenient summary of the dispersion of a set of data based on all values. In this manner, each value influences the mean deviation. A value that is close to the mean contributes little to the mean deviation. A value that is further away from the mean contributes more. With this measure, the presence of extremely large or extremely small values can be shown.

1.2.2 Variation and Standard Deviation

In calculating the mean deviation, we use the absolute values of the differences between data values and the mean as deviations because we need to make sure that positive deviations are not offset by negative deviations. Another way to avoid the offset caused by adding positive deviations to negative deviations is to square all deviations before summing them. The *variance* is one such measure. It can be calculated as

$$\sigma^2 = \frac{\sum_{i=1}^{n}(x_i - \overline{X})^2}{n},$$

where σ^2 is the variance. The i, n, and $\overline{X}$ are the same as those defined earlier. The equation for the variance actually calculates the average squared deviation of each value from the mean. While it is easier to understand it is not efficient in computation. A more computationally efficient formula for variance is

$$\sigma^2 = \frac{\sum_{i=1}^{n} x_i^2}{n} - \overline{X}^2.$$

This formula is more efficient because it minimizes the rounding error introduced by taking the differences and then squaring them.

Although variance measures dispersion in a data set, it is not commonly used because of its large numeric value. The deviations are squared before they are averaged. The process of squaring the deviations often leads to large numbers that cannot be compared directly to the original data values. As a remedy, the square root of the variance is often used to describe the dispersion of a data set. This measure is known as the *root mean square deviation*, or simply *standard deviation*. It can be calculated as

$$\sigma = \sqrt{\frac{\sum_{i=1}^{n}(x_i - \overline{X})^2}{n}}$$

or

$$\sigma = \sqrt{\frac{\sum_{i=1}^{n} x_i^2}{n} - \overline{X}^2}.$$

The standard deviation value is similar in numeric range to the data values. It is used more often than variance because taking the root of the squared deviation returns the magnitude of the value to that of the data set.

As an example, Table 1.5 shows the calculation of the standard deviation of the population densities from the 12 South American countries. For these 12 population densities, the mean is 43.916 (rounded to 44). The variance is 748. Therefore, the standard deviation is 27.35 because $\sqrt{748} = 27.35$.

Similarly, the variance for the population density values of all countries in the Americas is 372,443.36, and for the Central American countries it is 122,734.90. The standard deviations are 350.34 for the Central American countries and 610.28 for all countries in the Americas.

TABLE 1.5   Variance and Standard Deviation

Country	Population Density x	$x - \bar{X}$	$(x - \bar{X})^2$
Argentina	31	−13	169
Bolivia	18	−26	676
Brazil	46	2	4
Chile	48	4	16
Colombia	78	34	1156
Ecuador	108	64	4096
Guyana	9	−35	1225
Suriname	8	−36	1296
Paraguay	31	−13	169
Peru	49	5	25
Uruguay	45	1	1
Venezuela	56	12	144
$\sum$	527		8977
$\bar{X}$	44		748

The standard deviation has another useful property to help describe how values in a data set distribute. Statistically, the inclusion of data values in a value range bounded by standard deviations results in a well-known relationship if the distribution of the data closely resembles that of a normal distribution:

1. About 68% of the data values are within 1 standard deviation on either side of the mean, that is, values within an interval bounded by $\overline{X} - \sigma$ and $\overline{X} + \sigma$.
2. About 95% of the data values are within 2 standard deviations on either side of the mean, that is, values within an interval bounded by $\overline{X} - 2\sigma$ and $\overline{X} + 2\sigma$.
3. About 99% of the data values are within 3 standard deviation on either side of the mean, that is, values within an interval bounded by $\overline{X} - 3\sigma$ and $\overline{X} + 3\sigma$.

Similar to the calculation of the mean, a weighted variance and the associated weighted standard deviation can be derived from data representing observations grouped into classes. Adopting the same notations used before, the weighted variance is defined as

$$\sigma_w^2 = \frac{1}{\sum_{i=1}^{k} f_i} \left[\sum_{i=1}^{k} f_i(x_i - \overline{X}_w)^2 \right].$$

This intuitively meaningful formula also has its computational counterpart. For more efficient computation of the grouped variance, the following formula should be used:

$$\sigma_w^2 = \frac{1}{\sum_{i=1}^{k} f_i} \left[\sum_{i=1}^{k} f_i x_i^2 - \sum_{i=1}^{k} f_i(\overline{X}_w)^2 \right].$$

Then the standard deviation for the grouped data is the square root of the weighted variance.

The mean and the standard deviation describe where the center of a distribution is and how much dispersion a distribution has. Together they provide a sketch of the distribution as a basis for understanding a data set or comparing multiple data sets.

1.2.3 Skewness and Kurtosis

For a set of values, the mean gives its central tendency. The standard deviation suggests how much the values spread over the numeric range around the mean. There are also other characteristics of a numeric distribution that can be described by using additional measures. These include *skewness*, which measures the *directional bias* of a numeric distribution in reference to the mean, and *kurtosis*, which measures the *peakness* of a numeric distribution. Combining the mean, the standard deviation, the skewness, and the kurtosis, we have a set of descriptive statistics that can give rather detailed information about a given numeric distribution.

To understand how the skewness and kurtosis of a numeric distribution are calculated, it is necessary to discuss the concept of *frequency distribution*. The frequency distribution is often shown in a *histogram* in which the horizontal axis shows the numeric range of the data values and the vertical axis shows the frequency, that is, the number of values in each interval. Figure 1.4 shows five examples of frequency distributions with different levels of skewness and kurtosis. At the top is a symmetric distribution with low skewness and medium kurtosis. The two skewed distributions in the middle row show distributions with directional bias but low kurtosis. The two distributions in the bottom row show the

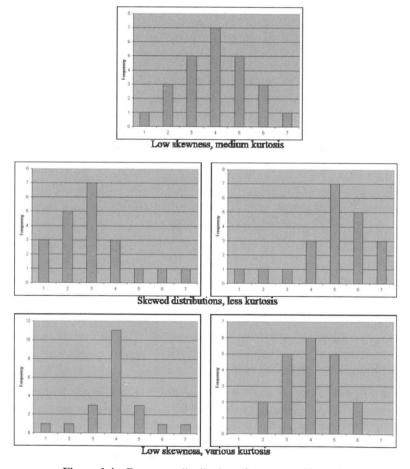

Figure 1.4   Frequency distribution: skewness and kurtosis.

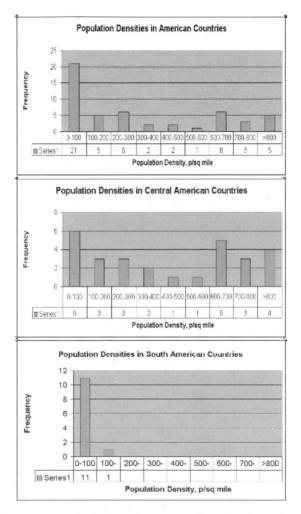

Figure 1.5   Frequency distribution of population densities of American countries.

difference between two kurtosis values. Figure 1.5 shows the frequency distributions of population density of the three America regions.

Skewness measures the extent to which the bulk of data values in a distribution are clustered to one side or the other side of the mean. When most values are less than the mean, the distribution is said to be *positively skewed*. Alternatively, a *negatively skewed* distribution is a distribution in which the bulk of the values are greater than the mean. Specifically, skewness can be calculated by

$$\text{Skewness} = \frac{\sum_{i=1}^{n}(x_i - \overline{X})^3}{n\sigma^3}.$$

where x, $\overline{X}$, σ, and n are the same as defined earlier. Notice that the measure of skewness is based on the cubic value of the deviations from the mean (or mean deviation) and the cubic value of the standard deviation, σ.

Because σ is positive, the denominator of the skewness formula is always positive. The numerator, however, can be positive or negative. If most of the values are smaller than the mean, the numerator will be negative and thus the distribution will be positively skewed. If most values are larger than the mean, the numerator will be positive and the skewness measure will be negative. The skewness of a symmetric distribution is 0 (zero).

Take the data set of population densities in South American countries as an example. Table 1.6 shows that

$$\sum_{i=1}^{n}(x_i - \overline{X})^3 = 191873.$$

Since σ is 27.35, as derived earlier, the skewness can be calculated as follows:

$$\text{Skewness} = \frac{\sum_{i=1}^{n}(x_i - \overline{X})^3}{n\sigma^3} = \frac{191{,}873}{12 \times 27.35^3} = \frac{191{,}873}{245{,}501} = 0.7816.$$

The distribution is thus moderately skewed in the positive direction, that is, countries with density levels higher than the mean are more frequent than countries with below-average density.

TABLE 1.6 Skewness and Kurtosis

Country	Population Density x	$x - \overline{X}$	$(x - \overline{X})^2$	$(x - \overline{X})^3$	$(x - \overline{X})^4$
Argentina	31	-13	169	-2197	28561
Bolivia	18	-26	676	-17576	456976
Brazil	46	2	4	8	16
Chile	48	4	16	64	256
Colombia	78	34	1156	39304	1336336
Ecuador	108	64	4096	262144	16777216
Guyana	9	-35	1225	-42875	1500625
Suriname	8	-36	1296	-46656	1679616
Paraguay	31	-13	169	-2197	28561
Peru	49	5	25	125	625
Uruguay	45	1	1	1	1
Venezuela	56	12	144	1728	20736
$\sum$	527		8977	191873	21829525
$\overline{X}$	44		748		

Skewness is most useful when it is used to compare distributions. For example, two distributions can have similar means and similar standard deviations, but their skewness can be very different if there are different directional biases.

With kurtosis, the extent to which values in a distribution are concentrated in one part of a frequency distribution can be measured. If the bulk of values in a distribution have a high degree of concentration, the distribution is said to be very *peaky*. Alternatively, a *flat* distribution is one without significant concentration of values in one part of the distribution.

The kurtosis is usually measured by the following equation:

$$\text{Kurtosis} = \frac{\sum_{i=1}^{n}(x_i - \overline{X})^4}{n\sigma^4} - 3.$$

The kurtosis is based on the fourth power of deviations of the values from their mean and the fourth power of the standard deviation, σ.

By subtracting 3 from the first part of the kurtosis equation, we structure the calculation of kurtosis so that a symmetrical, bell-shaped distribution has a value close to 0. In this way, a peaky distribution will have a positive kurtosis value and a flat distribution will have a negative kurtosis value.

Still using the population density values in South America, Table 1.6 gives $\sum_{i=1}^{n}(x_i - \overline{X})^4 = 21829525$ and $\sigma^2 = 748$. Therefore,

$$\begin{aligned}
\text{Kurtosis} &= \frac{\sum_{i=1}^{12}(x_i - \overline{X})^4}{n\sigma^4} - 3 \\
&= \frac{21,829,525}{12 \times 748^2} - 3 = 3.25 - 3 \\
&= 0.25
\end{aligned}$$

giving a distribution that is slightly peaky.

ArcView Notes When using ArcView to calculate descriptive statistics, the procedures are conducted in Table documents. There is a set of basic descriptive statistics users can derive by using the following steps:

1. Open the Table document from the Project window or open a theme table by selecting the menu item **Theme/Table** after adding a theme to a View document and have the theme active.
2. Click the title button of the field that contains the numerical values for which the statistics are needed.
3. Select the menu item **Field/Statistics** to have ArcView calculate the statistics, including sum (or total), count (i.e., *n*),

mean, maximum, minimum, range, variance, and standard deviation of the field.

While these basic descriptive statistics provided by the original ArcView are useful, they are not complete because the **Field/Statistics** does not produce skewness, kurtosis, or other values discussed in this chapter. To find these values, you can use a project file available on the companion website to this book that was developed to provide all descriptive statistics discussed in this chapter.

The project file, `ch1.apr`, which can be found in the `AVStat\Chapter1\Scripts\` directory in the archive downloaded from the support website, can be used to calculate additional statistics using the following steps:

1. Start a new project by selecting the menu item **File/Open Project** from the Project window.
2. Navigate to `\AVStat\Chapter1\Scripts\` and select `Ch1.apr`.
3. Add themes from your data directory to the view. You can add more than one theme, but make the theme for which you want to calculate additional statistics active.
4. Open the Table document of the active theme and click the title button of the field that contains the numerical values you

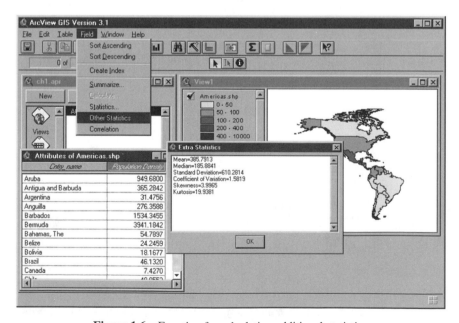

Figure 1.6   Function for calculating additional statistics.

want to use. Be sure that the field is a numeric field, not a string field.

5. Select the menu item **Field/Other Statistics** to have ArcView calculate additional statistics.

Figure 1.6 shows the location of the menu item and the results of calculating additional statistics, including skewness and kurtosis. Please note that results from the calculation using the **Field/Other Statistics** on data accompanied by the book will be different from the results reported in the text and in Tables 1.5–1.6. These discrepancies are rounding errors introduced when rounded values were used in the text and tables for illustrations.

1.3 RELATIONSHIP

The descriptive statistics discussed in the previous sections are useful for understanding and comparing how values distribute within one data set or between data sets. The mean, standard deviation, skewness, and kurtosis, although providing a basis for comparing different distributions, cannot measure the relationship between distributions quantitatively. To do so, we will need to apply the technique that this section discusses. This technique is based on the concept of *correlation*, which measures statistically the *direction* and the *strength* of the relationship between two sets of data or two variables describing a number of observations.

Given two counties in the same region where a similar tax code and similar developmental strategies have been applied, a comparison of their average family income figures will give us an impression of how well each county performs economically. If we consider subscribing to the concept that more spending on higher education will result in better economic progress, a look at the relationship between spending on higher education and some indicators of economic status will provide a potential answer. For this type of comparison, we typically measure how strongly the values of these two variables are related and the direction of their relationship.

The direction of the relationship of two variables is positive (or *direct*) if one of the values in a variable behaves similarly to another variable. For example, when the value of one variable for a particular observation is high, the value of the other variable for that observation is likely to be high. Alternatively, a *negative* (or *inverse*) relationship between two variables indicates that the value of one variable increases when the value of the other variable decreases. Of course, the stronger the relationship is, the more predictable this pattern will be.

In Figure 1.7, there are three diagrams that plot pairs of values as points in what are called *scatterplots*. In the top diagram, the relationship between the total length of motorways in the United Kingdom in 1993 is positively related to the total number of vehicles by region in 1991. Notice that the points show a pattern

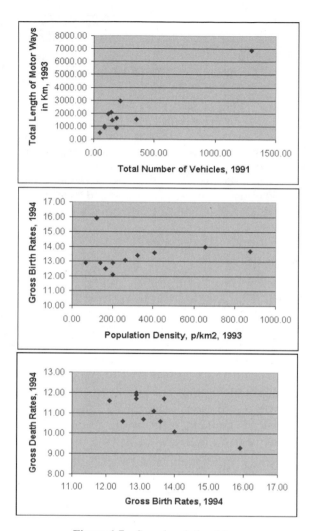

Figure 1.7 Sample relationships.

that runs from lower left to upper right. Taking any region as an example, it seems that a higher total number of vehicles in a region is often associated with a higher total length of motorways. In the lower diagram, the gross death rates and gross birth rates in regions of the United Kingdom in 1994 clearly show an inverse relationship. The pattern, as illustrated by the points in this scatterplot, indicates that a region with a high birth rate have a low death rate. The middle diagram shows the relationship between birth rate and population density. In this case, there does not seem to be a relationship at all; the trend as depicted by the points is flat, forming an almost horizontal line.

Beyond the direction of the relationship between two sets of data values, the strength of the relationship can be estimated quantitatively, and of course visually, from the scatterplots. In Figure 1.7, the top plot indicates a stronger relationship than the relationship described by the bottom plot. This is because the point distribution in the top plot is less scattered than that of the bottom plot.

Quantitatively, we can measure the direction and the strength of the relationship between two sets of data values by calculating the correlation coefficient. In this section, we will discuss only what is known as the *product-moment correlation coefficient* (or *Pearson's correlation coefficient*). This coefficient works best for *interval/ratio scale* data. For data measured at nominal or ordinal scales, other coefficients (χ^2 and Spearman's rank coefficient) should be applied.

The Pearson's correlation coefficient, r, between two variables, x_i and y_i, $i = 1, 2, \ldots, n$, can be calculated by

$$r = \frac{\sum_{i=1}^{n}(x_i - \overline{X})(y_i - \overline{Y})}{(n-1)S_x S_y},$$

where S_x and S_y are the standard deviations of x and y, respectively.

The numerator is essentially a covariance, indicating how the two variables, x and y, vary together. Each x_i and y_i is compared with its corresponding mean. If both x_i and y_i are below their means, the product of the two negatives will be a positive value. Similarly, if both are large, the product will be positive, indicating a positive correlation. If x_i is larger than the mean but y_i is smaller than the mean, the product will be negative, indicating an inverse relationship. The sum of all these covariations reflects the overall direction and strength of the relationship. For computational efficiency, we prefer to use the formula

$$r = \frac{\dfrac{\sum_{i=1}^{n} x_i y_i}{n} - \overline{X} \bullet \overline{Y}}{S_x S_y},$$

where $\overline{X}$ and $\overline{Y}$ are means for x and y, respectively, and S_x and S_y are standard deviations for x and y, respectively, defined as

$$S_x = \sqrt{\frac{\sum_{i=1}^{n} x^2}{n} - \overline{X}^2}$$

and

$$S_y = \sqrt{\frac{\sum_{i=1}^{n} y^2}{n} - \overline{Y}^2}.$$

This coefficient is structured so that the sign of the value of r indicates the direction of the relationship as:

$r > 0$ when the relationship between the two variables is a direct (or positive) one,

$r < 0$ when the relationship between the two variables is an inverse (or negative) one, and

$r \approx 0$ when there is no relationship between the two variables.

The absolute value of r indicates the strength of the relationship with a numeric range of:

$r = -1$ for the strongest or perfectly inverse relationship and

$r = 1$ for the strongest or perfectly direct relationship.

TABLE 1.7   Data for Pearson's Product-Moment Correlation Coefficient

Regions $(n = 11)$	Total Length of Motorways x	Total No. of Vehicles y	x^2	y^2	xy
North	152	923	23,104	851,929	140,296
Yorkshire and Humberside	289	1629	83,521	2,653,641	470,781
East Midlands	185	1438	34,225	2,067,844	266,030
East Anglia	22	890	484	792,100	19,580
South East	919	6893	844,561	47,513,449	6,334,667
South West	302	1960	91,204	3,841,600	591,920
West Midlands	378	2066	142,884	4,268,356	780,948
North West	486	2945	236,196	8,673,025	1,431,270
Wales	120	979	14,400	958,441	117,480
Scotland	285	1545	81,225	2,387,025	440,325
Northern Ireland	113	483	12,769	233,289	54,579

$n = 11$

$\sum x = 3251$ $\qquad$ $\sum y = 21751$

$\overline{X} = 295.55$ $\qquad$ $\overline{Y} = 1,977.36$

$\sum x^2 = 1,564,573$ $\qquad$ $\sum y^2 = 74,240,699$

$\overline{X}^2 = 87,349.81$ $\qquad$ $\overline{Y}^2 = 3,909,952.57$

$S_x = \sqrt{\frac{1,564,753}{11} - 87,347.12}$ $\qquad$ $S_y = \sqrt{\frac{74,240,699}{11} - 3,939,952.57}$

$= 234.31$ $\qquad$ $= 1,684.99$

$\sum xy = 10,647,876$

$$r = \frac{\frac{10,647,876}{11} - 295.55 \times 1,977.36}{234.31 \times 1,684.99} = \frac{967,988.73 - 584,408.75}{394,810}$$

$$= \frac{383,579.98}{394,810} = 0.97$$

Table 1.7 shows an example of the calculation of Pearson's product-moment correlation coefficient. The two variables are:

x: Total number of private vehicles (except tractors and motorcycles) in 1991

y: Total length of motorways in kilometers in 1993.

The resulting correlation coefficient 0.97, indicating a very strong, positive relationship between the two variables. Specifically, a region that has a larger number of vehicles also has a longer total motorway.

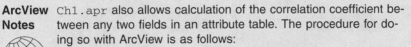

ArcView `Ch1.apr` also allows calculation of the correlation coefficient be-
Notes tween any two fields in an attribute table. The procedure for do-
ing so with ArcView is as follows:

1. Start ArcView and select the menu item **File/Open Project** from the Project window.
2. Navigate to the directory where you might have copied the project file to open `Ch1.apr` by highlighting it and clicking the **OK** button.
3. Use the **Add Theme** button to bring in any theme that contains the attribute fields you want to calculate the correlation coefficients. Make sure that the theme is the active theme and then open its attribute table by clicking the **Table** button under the **Theme** menu.
4. Select the **Field/Correlation** menu item.
5. From the pop-up dialogue box, select the first variable from the drop-down list of attributes. Click the **OK** button to proceed.
6. From the next pop-up dialogue box, select the second variable and then click the **OK** button to calculate the correlation coefficient.

As shown in Figures 1.8 and 1.9, the steps for calculating a correlation coefficient are in fact quite straightforward. A word of caution: this procedure should not be applied to fields in an attribute table that contain either nominal or ordinal data or strings. Note that the correlation function is also accessible under the **Statistics/Correlation** menu item when the View is active.

1.4 TREND

The previous section focused on the technique for measuring the direction and strength of the relationship between two variables. In this section, we will discuss the technique for measuring the trend, as shown by the relationship between two

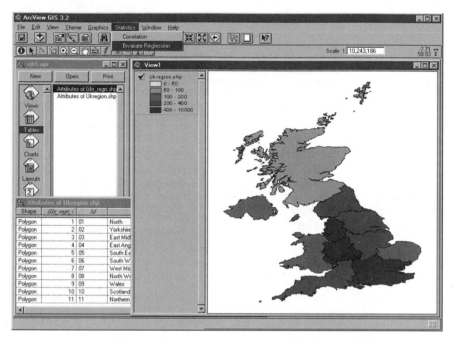

Figure 1.8 Statistical function of correlation.

Selecting the first variable:

Selecting the second variable:

The resulting correlation coefficient:

Figure 1.9 Steps for calculating the correlation coefficient.

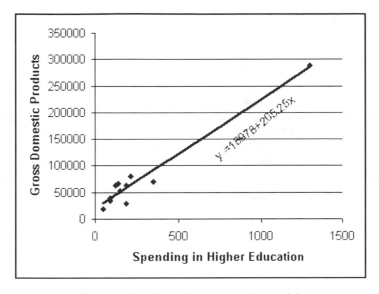

Figure 1.10   Simple linear regression model.

variables. The technique shows the *dependence* of one variable on another. Going beyond knowing the strength and direction of the relationship, the technique for measuring the trend allows us to estimate the likely value of one variable based on a known value of another variable. This technique is known as the *regression* model. Although the regression model does not imply a *causal relationship*, it provides the information necessary for the *prediction* of one variable by another.

To illustrate what measuring the trend of the relationship between two variables means, Figure 1.10 shows the relationship between expenditure on higher education in 1993 and gross domestic product (GDP) (in million European Community Units (ECU)) in 1993 by region in the United Kingdom. It can be seen that there is a positive, strong correlation between the two variables. For regions where GDP was high, spending on higher education was also high. Alternatively, the relationship shows that regions with less spending on higher education are also regions with lower GDP. The straight line running between the data points is the regression line. Because this trend line is a linear one, this type of regression model is also called *simple linear regression* or *bivariate regression* because it involves only two variables.

The simple linear regression model is the simplest form of regression model. It is generally represented as

$$Y = a + bX,$$

where

    Y is the dependent variable,
    X is the independent variable,

a is the intercept, and

b is the slope of the trend line.

The variable that is used to predict another variable is the independent variable (X). The variable that is predicted by another variable is the dependent variable (Y). The intercept is the value of the dependent variable when the value of the independent variable is zero or the value of Y when the regression line intersects the y-axis. Finally, the slope is the rate of change of the dependent variable's value as a per unit change of the independent variable.

The procedure for finding the trend line is to *fit a regression line* among all data points. Specifically, the procedure is to find the value of *a* (intercept) and the value of *b* (slope) in the regression line $Y = a + bX$:

$$b = \frac{\sum xy - n\overline{X}\overline{Y}}{\sum x^2 - n\overline{X}^2}$$

and

$$a = \overline{Y} - b\overline{X},$$

where $\sum xy$ is the sum of the products $x_i y_i$, $i = 1, 2, \ldots, n$; $\sum x^2$ is the sum of squared values of the independent variables; and $\overline{X}$ and $\overline{Y}$ are the means of the independent and dependent variables, respectively.

As an example, Table 1.8 shows the steps for calculating the values of *a* and *b* to construct the trend line in a simple linear regression model. The resulting regression model is

$$\hat{y} = 18,977.96 + 205.25x.$$

The intercept is 18,977.96 and the slope is 205.25. With this model, we can calculate a set of estimated values for the dependent variable using the values we have for the independent variable. The results of this estimation are listed in Table 1.9. As shown in this table, there are some deviations between the observed values and the estimated values of the dependent variable. The deviations between the observed and predicted values are known as *residuals*. Ideally, a perfect regression model will have zero residuals. The larger the residuals, the less powerful the regression model is. When the residuals are small, we typically say that the regression line is a *good fit*.

Since regression models are not always equally powerful, how do we know if a regression model is a good enough fit of the data? To answer this question, we need to calculate a *coefficient of determination*, usually denoted as r^2. The coefficient of determination is the ratio between the variance of the predicted values of the dependent variable and the variance of the observed values of the

TABLE 1.8   Data for Simple Linear Regression Model

Name	1993 GDP in Million ECU y	1993 Expenditure on Higher Education x	x^2	xy
North	38,494.4	92.3	8,519.29	3,553,033.12
Yorkshire and Humberside	63,701.6	191.0	36,481.0	12,167,005.60
East Midlands	52,937.1	155.1	24,056.01	8,210,544.21
East Anglia	29,552.9	188.5	35,532.25	5,570,721.65
South East	288,479.1	1,297.5	1,683,506.25	374,301,632.25
South West	62,739.0	123.1	15,153.61	7,723,170.90
West Midlands	67,161.5	142.3	20,249.29	9,557,081.45
North West	80,029.7	219.2	48,048.64	17,542,510.24
Wales	34,028.7	91.0	8,281.0	3,096,611.70
Scotland	69,601.0	355.1	126,096.01	24,715,315.10
Northern Ireland	18,033.3	48.7	2,371.69	878,221.71

$$\sum y = 804,758.30 \qquad \sum y = 2,903.80$$
$$n = 10 \qquad n = 10$$
$$\overline{Y} = 73,159.85 \qquad \overline{X} = 263.98$$
$$\sum xy = 467,315,847.93 \qquad \sum x^2 = 2,008,295.04$$

$$b = \frac{467,315,847.93 - 11(263.98)(73,159.85)}{2,008,295.04 - (11)(263.98)(263.98)} = \frac{254,875,738.7}{1,241,755.2} = 205.25$$

$$a = 73,159.85 - (205.5)(263.98) = 18,977.96$$

$$\hat{y} = 18,977.96 + 205.25x$$

dependent variable. Specifically, the coefficient of determination can be calculated as

$$r^2 = \frac{S_{\hat{y}}^2}{S_y^2},$$

where $S_{\hat{y}}^2$ is the variance of the predicted values of the dependent variable, also known as *regression variance*, and S_y^2 is the variance of the observed values of the dependent variable, also known as *total variance*.

In the lower part of Table 1.9, the coefficient of determination is calculated as $r^2 = 0.9799$. Converting this ratio to a percentage, we can say that 97.99% of the variance of the dependent variable is accounted for or captured by the regression. Consequently, the higher the r^2 value is, the better the regression model's fit to the data values. Also note that the square root of the coefficient of determination is r, which is the Pearson's product-moment correlation coefficient discussed in the previous section.

TABLE 1.9   Regression Model: Observed Values, Predicted Values, and Residuals

Regions	x	Observed y	Predicted $\hat{y}$	Residuals $y - \hat{y}$
North	92.3	38,494.4	39,722.54	−1,228.14
Yorkshire and Humberside	191.0	63,701.6	58,180.71	5,520.89
East Midlands	155.1	52,937.1	50,812.24	2,124.87
East Anglia	188.5	29,552.9	57,667.59	−28,114.69
South East	1,297.5	288,479.1	285,289.84	3,189.26
South West	123.1	62,739.0	44,244.24	18,494.77
West Midlands	142.3	67,161.5	48,185.04	18,976.47
North West	219.2	80,029.7	63,968.76	16,060.94
Wales	91.0	34,028.7	37,655.71	−3,627.01
Scotland	355.1	69,601.0	91,862.24	−22,261.24
Northern Ireland	48.7	18,033.3	28,973.64	−10,940.34

$$\sum y^2 = 1.13615 \times 10^{11}$$

$$\sum \hat{y}^2 = 1.11328 \times 10^{11}$$

Total variance: $S_y^2 = \dfrac{\sum y^2}{n} = 1.0329 \times 10^{10}$

Regression variance: $S_{\hat{y}}^2 = \dfrac{\sum \hat{y}^2}{n} = 1.0121 \times 10^{10}$

Coefficient of determination: $r^2 = \dfrac{1.0121 \times 10^{10}}{1.0329 \times 10^{10}} = 0.9799$

ArcView Notes

Similar to the calculation of the correlation coefficient, the project file, Ch1.apr, can be used to load themes and to calculate the simple linear regression model. The menu item for calculating regression can be found in the Statistics menu for View documents.

To use Ch1.apr for calculating a regression model, do the following:

1. Start ArcView and use **File/Open Project** to load Ch1.apr, as before.

2. Start a new View document and use the **Add Theme** button to bring in themes you want to use.

3. From the menu of the View document, choose **Statistics/ Bivariate Regression** to initiate the procedure.

4. In the pop-up dialog box, choose the field for the independent variable and then click **OK** to proceed.

5. In the next pop-up dialog box, choose the field for the dependent variable and then click **OK** to proceed.

6. The results are shown in another pop-up dialog box, including the values for the coefficient of determination, intercept, and slope of the regression model.

7. After you click the **OK** button in the dialog box that shows the results, another dialog box appears to ask if a scatterplot of the data should be created. Click **Yes** to create it or **No** to end the calculation. When the scatterplot appears, you can expand the window horizontally to extend the plot in the X direction.

As an example, Figure 1.11a shows the **Statistics/Bivariate Regression** menu item. Figure 1.11b shows the steps in performing the calculation of bivariate regression and its results.

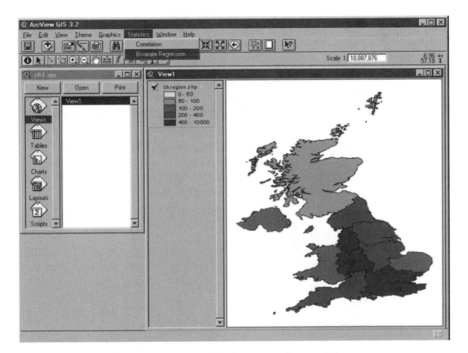

Figure 1.11a The bivariate regression menu item.

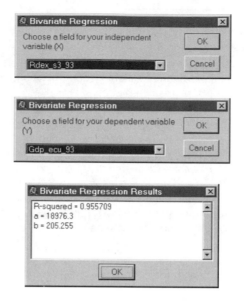

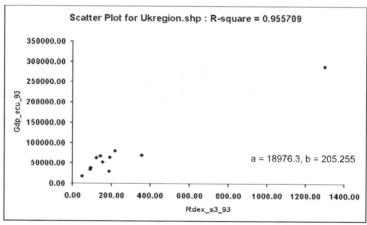

Figure 1.11b Steps and results of bivariate regression.

CHAPTER 2

POINT DESCRIPTORS

In the Introduction and in Chapter 1, we discussed that spatial data consist of cartographic data (which describe the locational and geometric characteristics of features) and attributes data (which provide meanings to the geographic features). In Chapter 1, we focused on the analysis of attribute data exclusively, and thus the analysis was aspatial. Starting with this chapter, we will discuss statistical tools that have been implemented in GIS and have been especially designed to analyze only locational information or locational information together with attribute information. In this chapter, the locational information of points will be used in several descriptive geostatistics or centrographic measures to analyze point distribution.

2.1 THE NATURE OF POINT FEATURES

With vector GIS, *geographic features* are represented geometrically by points, lines, or polygons in a two-dimensional space. Geographic features that occupy very little or no areal extent at a given scale (the scale of study) are often represented as *points*. In a vector GIS database, linear features are best described as *lines*, and regions or areal units are often structured as *polygons*. As a whole, we refer to points, lines, and polygons as *geographic objects* since they represent geographic features.

While geographic objects can be conveniently represented as points, lines, and polygons, the relationship between geographic objects and the geographic features they represent is not always fixed. Scales often determine how geographic features are represented. For example, a house on a city map is only a point, but it becomes a polygon when the floor plan of the house is plotted on a map. Similarly, the City of Cleveland, Ohio, is represented by a large polygon that occupies an entire map sheet. In this way, the details of the city's street network and other facilities are shown on a large-scale map (large because the ratio may be 1:2,000 or larger). Alternatively, Cleveland is often represented only as a point on a small-scale map (small because the ratio may be 1:1,000,000 or smaller) that identifies all of the major cities in the world.

The degree of abstraction also affects how various geographic objects represent geographic features. This is because points can be used not only to represent physical geographic features such those described above, but also to describe the locations of events and incidences such as disease occurrences or even traffic accidents. In these cases, points do not represent real geographic features, but just locations of the events. Furthermore, for transportation modeling, urban analysis, or location-allocation analysis, areal units with significant spatial extents are often abstractly represented by points (such as centroids) to accommodate specific data structures, as required by the analytic algorithms.

In this chapter, we will focus on point features. Points are defined by coordinates. Depending on the coordinate system and the geographic projections, points on a map can be defined by a pair of latitude and longitude measures, x and y coordinates, easting and northing, and so on. On a small-scale map that covers a large areal extent but shows few geographic details, points can be used to identify locations of cities, towns, tourist attractions, and so on. On a large-scale map, points may represent historical landmarks, street lights, trees, wells, or houses.

While points on any map are all simple geometric objects that are defined by their coordinates, the attributes associated with these points provide specifics to differentiate them according to the characteristics emphasized. Consider a map showing all residential water wells in a city; the points will all look alike except for their locations. If attributes such as owners' names, depths, dates dug, or dates of last water testing were added to the database, more meaningful maps could be created to show spatial variation of the wells according to any of the attributes.

Individually, points may represent the locations of geographic features. The associated attributes help to describe each point's unique characteristics. The description of spatial relationship between individual points, however, requires the application of some of the spatial statistics described in this chapter. Specifically, we will discuss ways to determine where points are concentrated, as described by their locations or weighted by a given attribute. We will also examine how to measure the degree of dispersion in a set of points. This set of tools is also known as centrographic measures (Kellerman, 1981).

The spatial statistical methods to be discussed in this chapter are appropriate for points that represent various types of geographic features in the real world. A word of caution is needed here: the accuracy of locational information and its associated attribute values must be considered carefully. This is because the reliability and usefulness of any inference that results from analyzing the points are often affected by the quality of the data.

Point data obtained from maps may contain cartographic generalizations or locational inaccuracy. On a small-scale map, a point may represent a city whose actual areal extent is a certain number of square miles, while another point may represent a historical landmark or the location of threatened plant species that occupy only several square inches on the ground. Comparing these points directly, however carefully performed, would be like comparing oranges with apples. Point locations derived from calculating or summarizing other point data can be especially sensitive to the quality of input data because the inaccuracy of input data will be propagated during computation, so that the results are of little value due to the magnitude of the inaccuracy.

For this reason, we urge spatial analysts to be sensitive to the scale of a spatial database and the quality of the data used in the analysis. Statistical methods can be very useful when they are used correctly, but they can be very misleading and deceiving when used inappropriately or carelessly.

2.2   CENTRAL TENDENCY OF POINT DISTRIBUTIONS

In classical statistics, the conventional approach to summarizing a set of values (or numerical observations) is to calculate the measure of central tendency. The central tendency of a set of values gives some indication of the *average* value as their representative. The average family income of a neighborhood can give an out-of-towner a quick impression of the economic status or living style of the neighborhood. If you plan to visit an Asian country for Christmas, it would be useful to know the average temperature in December there so that you know what clothing to take.

When comparing multiple sets of numerical values, the concept of average is particularly useful. Educators can use average scores of state proficiency tests between elementary schools to see how schools compare with one another. Comparing the harvests from farms using a new fertilizer with the harvests from farms without it provides a good basis for judging the effectiveness of the fertilizer. In these and many other similar settings, the central tendency furnishes summary information of a set of values that would otherwise be difficult to comprehend.

Given a set of values, $x_i, i = 1, 2, \ldots, n$, measures of central tendency include the mean, weighted mean, and median. The mean, $\overline{X}$, is simply the arithmetic average of all values as

$$\overline{X} = \frac{\sum_{i=1}^{n} x_i}{n}.$$

What if the observation values in a data set do not carry the same level of importance? Obviously, the measure of central tendency will not be simply the arithmetic mean. In that case, each value, x_i, in the data set will first be multiplied by its associated weight, w_i. The sum of the weighted values is then divided by the sum of the weights to obtain the *weighted mean*:

$$\overline{X}_W = \frac{\sum_{i=1}^{n} x_i w_i}{\sum_{i=1}^{n} w_i}.$$

Another measure of central tendency in classical statistics is the *median*. These two measures of central tendency were discussed in Chapter 1.

When dealing with a data set that contains observations distributing over space, one can extend the concept of *average* in classical statistics to the concept of *center*, as a measure of spatial central tendency. Because geographical features have spatial references in a two-dimensional space, the measure of central tendency needs to incorporate coordinates that define the locations of the features or objects. Central tendency in the spatial context will be the mean center, the weighted mean center, or the median center of a spatial point distribution.

There are several ways in which the position of such centers can be calculated; each gives different results based on how the data space is organized. Different definitions of the extent of the study area, distortions caused by different map projection systems, or even different map scales at which data were collected often lead to different results. It is important to realize that there is no one *correct* way of finding the center of a spatial distribution. There are appropriate methods for use in various settings, but there is probably no single correct method suitable for all situations. Therefore, the interpretation of the result of calculating the center of a spatial distribution can only be determined by the nature of the problem.

To describe a point distribution, we will discuss a series of point descriptors in this chapter. For central tendency, mean centers, weighted mean centers, and median centers provide a good summary of how a point set distributes. For the extent of dispersion, standard distances and the standard ellipse measure the spatial variation and orientation of a point distribution.

2.2.1 Mean Center

The *mean center*, or spatial mean, gives the average location of a set of points. Points may represents water wells, houses, power poles in a residential subdivision, or locations where landslides occurred in a region in the past. As long as a location can be defined, even with little or no areal extent, it can be represented as a point in a spatial database. Whatever the points in a spatial database represent, each point, p_i, may be defined operationally by a pair of coordinates, (x_i, y_i), for its location in a two-dimensional space.

The coordinate system that defines the location of points can be quite arbitrary. Geographers have devised various map projections and their associated coordinate systems, so the locations of points in space can be referred to by their latitude/longitude, easting/northing, or other forms of coordinates. When working with known coordinate systems, the location of a point is relatively easy to define or even to measure from maps. There are, however, many situations requiring the use of coordinate systems with an arbitrary origin as the reference point. Arbitrary coordinate systems are often created for small local studies or for quick estimation projects. In those cases, the coordinate system needs to be carefully structured so that (1) it orients to a proper direction for the project, (2) it situates with a proper origin, and (3) it uses suitable measurement units. For more detailed discussion of these selections, interested readers may refer to the monograph by

TABLE 2.1   Ohio Cities from the Top 125 U.S. Cities and Their Mean Center

City Name	Longitude in Degrees (x)	Latitude in Degrees (y)
Akron	-81.5215	41.0804
Cincinnati	-84.5060	39.1398
Cleveland	-81.6785	41.4797
Columbus	-82.9874	39.9889
Dayton	-84.1974	39.7791
$n = 5$	$\sum x = -414.8908$	$\sum y = 201.4679$
	$\overline{x}_{mc} = \dfrac{-414.8908}{5} = -82.9782$	$\overline{y}_{mc} = \dfrac{201.4679}{5} = 40.2936$

Monmonier (1993). All these issues have to be taken into account so that the resulting mean center will approximate its most appropriate location.

With a coordinate system defined, the mean center can be found easily by calculating the mean of the x coordinates (eastings) and the mean of the y coordinates (northings). These two mean coordinates define the location of the mean center as

$$(\overline{x}_{mc}, \overline{y}_{mc}) = \left(\frac{\sum_{i=1}^{n} x_i}{n}, \frac{\sum_{i=1}^{n} y_i}{n} \right),$$

where

$\overline{x}_{mc}, \overline{y}_{mc}$ are coordinates of the mean center,

x_i, y_i are coordinates of point i, and

n is the number of points.

As an example, Table 2.1 lists the coordinates of 5 cities in Ohio that are among the 125 largest U.S. cities. Their locations are shown in Figure 2.1. The calculation in Table 2.1 shows that the mean center of the five cities is located at -82.9782, 40.2936 or 82.9782W, 40.2936N. The star in Figure 2.1 identifies the location of the calculated mean center. Since this mean center is defined by the mean of x coordinates and the mean of y coordinates, it is located at the geometric center of the five cities, as expected. What it represents is the center of gravity of a spatial distribution formed by the five cities.

2.2.2 Weighted Mean Center

There are situations in which the calculation of mean centers needs to consider more than just the location of points in the spatial distribution. The importance of individual points in a distribution is not always equal. In calculating the spatial mean among a set of cities, the mean center may give a more realistic picture of the central tendency if the mean center is weighted by the population counts

Figure 2.1   Five Ohio largest 125 U.S. cities and their mean center with standard distance.

of these cities. The mean center is pulled closer to a city if the city's population is larger than the populations of the other cities being considered. Similarly, a *weighted mean center* provides a better description of the central tendency than a mean center when points or locations have different frequencies or occurrences of the phenomenon studied. Given points representing the locations of endangered plant species, it makes more sense to calculate their mean center by using the sizes of plant communities at each location as weight of the points.

The weighted mean center of a distribution can be found by multiplying the x and y coordinates of each point by the weights assigned to them. The mean of the weighted x coordinates and the mean of the weighted y coordinates define the position of the weighted mean center.

The equation for the weighted mean center is

$$(\bar{x}_{\text{wmc}}, \bar{y}_{\text{wmc}}) = \left(\frac{\sum_{i=1}^{n} w_i x_i}{\sum_{i=1}^{n} w_i}, \frac{\sum_{i=1}^{n} w_i y_i}{\sum_{i=1}^{n} w_i} \right),$$

where

$\bar{x}_{\text{wmc}}, \bar{y}_{\text{wmc}}$ defines the weighted mean center,

and

w_i is the weight at point p_i.

TABLE 2.2   Ohio Cities from the Top 125 U.S. Cities and Their Weighted Mean Center

City Name	Longitude in Degrees x	Latitude in Degrees y	Population in 1990 p
Akron	−81.5215	41.0804	223,019
Cincinnati	−84.5060	39.1398	364,040
Cleveland	−81.6785	41.4797	505,616
Columbus	−82.9874	39.9889	632,910
Dayton	−84.1974	39.7791	182,044
Sum	$\sum x = -414.8908$	$\sum y = 201.4679$	$\sum p = 1,907,629$

City Name	Longitude × Population $x \times p$	Latitude × Population $y \times p$
Akron	−18,180,843.41	9,161,709.73
Cincinnati	−30,763,564.24	14,248,452.79
Cleveland	−41,297,956.46	20,972,800.00
Columbus	−52,523,555.33	2,530,937.47
Dayton	−15,327,631.49	7,241,546.48
Sum	$\sum xp = -158,093,550.90$	$\sum yp = 76,933,883.70$
$n = 5$	$\sum xp = -158,093,550.9$	$\sum yp = 76,933,883.7$

$$\bar{x}_{\text{wmc}} = \frac{\sum xp}{\sum p} = \frac{-158,093,550.9}{1,907,629} \qquad \bar{y}_{\text{wmc}} = \frac{\sum yp}{\sum p} = \frac{76,933,883.7}{1,907,629}$$

$$= -82.87 \qquad\qquad\qquad = 40.33$$

In the case of the 5 largest 125 U.S. cities in Ohio, the mean center will be shifted toward the Cleveland-Akron metropolitan area if the population sizes are used as weights for the 5 Ohio cities. To calculate this weighted mean center, Table 2.2 lists the coordinates as those of Table 2.1 and population counts in cities in 1990.

The result shown in Table 2.2 gives the location of the weighted mean center as −82.87, 40.33, representing a shift toward the northeast direction from the mean center calculated in Section 2.2.1.

ArcView Notes   In calculating the mean center, the major inputs are the x and y coordinates for the unweighted case. The additional input will be the weights for the weighted mean. The weights are usually provided as an attribute (such as counts of incidence or the number of people) of that location. The x and y coordinates usually are not explicitly recorded in ArcView. They must be extracted either by using the **Field Calculator** with **.GetX** and **.GetY** requests issued to the point shape objects or by running the Avenue sample script, `addxycoo.ave`. Either approach will put the x and y coordinates into the attribute table associated with the point theme in ArcView. To use the **Field Calculator**, first add a theme to a View. Open the associated feature table under the Theme mean. Then choose **Start Editing** on the table under the Table menu. Under **Edit**, choose **Add Field** to add a field for the x-coordinate readings and a field for the y-coordinate readings.

After adding the two new fields to the Table, click on the new field for the x-coordinate, then go to **Field/Calculate** menu item. The Field Calculator will appear. It will ask for a formula for the x-coordinate. Double-click on the [Shape] field in the list of fields to include the shape field in the formula. Then type `.GetX`. Be sure to include the period. Then click **OK**. ArcView will extract the x-coordinate of all points and put the values in the new field. Repeat the same process for the y-coordinate but use `.GetY`. Afterward, remember to save the result by choosing the **Table/Stop Editing** menu item.

If one prefers to use the sample script `addxycoo.ave`, first open a new script window by double-clicking on the script icon in the Project window. Then go to **Script/Load Text File** to locate the sample script. Usually, it is under the subdirectory for ArcView\samples\scripts. Load the script into the script window. Go to **Script/Compile** to compile it before **Run**.

Similarly, to calculate the mean center in ArcView, we can use either the existing functions of ArcView or Avenue. For the unweighted case, the mean center is just the average of the x and y coordinates for all points. Therefore, using the **Statistics** function under the **Field** menu in **Table** documents, a set of statistics including the mean of the chosen coordinate (x or y) will be provided.

For the weighted case, the x and y coordinates have to be weighted by the appropriate attribute. First, two new columns (one for weighted x and another for weighted y) have to be created (follow the steps above to start editing a table and adding new fields). The two columns are then multiplied by their corresponding weights, as described in the equations above using the **Field Calculator**. Then the weighted mean center is obtained by using the **Statistics** function under the **Field** menu. Please note, however, that the results have to be recorded manually and cannot be displayed graphically.

If you are using Avenue to calculate mean centers, the process of extracting the coordinates can be coded into the Avenue script. In that case, the coordinates will not need to be stored in the attribute table. Therefore, the attribute table does not need to be modified. This book has included a project file, Ch2.Apr, in the accompanying website. This file can be loaded into ArcView as a project (when loading ArcView, use the menu item **File** and then **Open Project** to direct ArcView to the appropriate directory to open this project file). All the geostatistical tools described in this chapter are included in this project file. The layout and the ArcView user interface of that customized project file are shown in Figure 2.2. In contrast to an ordinary ArcView project interface, this project has an additional menu item, **Spatial Analysis**. Under this menu, you can find items for **Spatial Mean**, **Standard Distance**, **Spatial Median**, and **Standard Deviational Ellipse** calculations. This project file, however, is encrypted to prevent

Figure 2.2 Customized ArcView user interface.

novice users from accidentally changing the content of the scripts.

The script for the mean center calculation includes the procedures for extracting the coordinates of point features and the mean center calculation. The script can handle both unweighted and weighted cases, allowing the user to select an attribute from the attribute table of a point theme as the weight. In addition, the script will draw the result (a point) on the associated View document to show the location of the mean center. Finally, if the user wishes to store the location of mean centers for future analysis, there is an option for creating a shapefile of the mean center to be used later.

2.2.3   Median Center

Analogous to classical descriptive statistics of central tendency, the concept of the median of a set of values can be extended to the *median center* of a set of points. But the median in a geographical space cannot be defined precisely. According to Ebdon (1988), the median center of a set of points is defined differently in different parts of the world. In the British tradition, given a set of points, a median center is the center upon which the study region is divided into four quadrants, each containing the same number of points. However, there can be more than one median center dividing the study area into four parts with equal numbers of points if there is sufficient space between points close to the center of the distribution. As this method leaves too much ambiguity, it has not been used extensively.

As used in North America, the concept of median center is the center of minimum travel. That is, the total distance from the median center to each of the points in the region is the minimum. In other words, any location other than the median center will yield a total distance larger than the one using the median center. Mathematically, median center, (u, v), satisfies the following objective function:

$$\text{Min} \sum_{i=1}^{n} \sqrt{(x_i - u)^2 + (y_i - v)^2},$$

where x_i and y_i are the x and y coordinates, respectively, of point p_i. If there are weights attached to the points, a weighted median center can be derived accordingly:

$$\text{Min} \sum_{i=1}^{n} f_i \sqrt{(x_i - u)^2 + (y_i - v)^2}.$$

Please note that the weights, f_i for p_i, can be positive or negative values to reflect the pulling or pushing effects of points to the location of the median center.

To derive the median center, an iterative procedure can be used to explore and to search for the location that satisfies the above objective function. The procedure is as follows:

1. Use the mean center as the initial location in searching for the median center. This is essentially setting (u_0, v_0) equal to (x_{mc}, y_{mc}).
2. In each iteration, t, find a new location for the median center, (u_t, v_t), by

$$u_t = \frac{\sum f_i x_i \Big/ \sqrt{(x_i - u_{t-1})^2 + (y_i - v_{t-1})^2}}{\sum f_i \Big/ \sqrt{(x_i - u_{t-1})^2 + (y_i - v_{t-1})^2}}$$

and

$$v_t = \frac{\sum f_i y_i \Big/ \sqrt{(x_i - u_{t-1})^2 + (y_i - v_{t-1})^2}}{\sum f_i \Big/ \sqrt{(x_i - u_{t-1})^2 + (y_i - v_{t-1})^2}}.$$

3. Repeat step 2 to derive new locations for the median center until the distance between (u_t, v_t) and (u_{t-1}, v_{t-1}) is less than a threshold defined a priori.

ArcView Notes The Avenue script for median centers, as incorporated into the project file `Ch2.apr`, is an extension of the script for the mean center. This is because the mean center is used as the initial location. The script asks the user for a threshold distance to control the termination of the script. Using the mean center as the

Figure 2.3 Spatial median.

initial location, the script goes through the iterative process to search for the median center. Figure 2.3 shows the median center, which is quite far from the spatial mean. Please note that if the coordinates of points are in degrees of longitude and latitude (Map Units, as defined in ArcView), the threshold distance is also defined in that metric scale (degrees).

2.3 DISPERSION OF POINT DISTRIBUTIONS

Similar to using measures such as standard deviations to assist an analyst in understanding a distribution of numeric values, standard distances or standard ellipses have been used to describe how a set of points disperses around a mean center. These are useful tools because they can be used in very intuitive ways. The more dispersed a set of points is around a mean center, the longer the standard distance and the larger the standard ellipse it will have.

Given a set of n data values, x_i, $i = 1, \ldots, n$, the *standard deviation*, S, can be computed as

$$S = \sqrt{\frac{\sum_{i=1}^{n}(x_i - \overline{x})^2}{n}},$$

where $\overline{x}$ is the mean of all values. The standard deviation is literally the square root of the average squared deviation from the mean.

2.3.1 Standard Distance

Standard distance is the spatial analogy of standard deviation in classical statistics. While standard deviation indicates how observations deviate from the mean, standard distance indicates how points in a distribution deviate from the mean center. Standard deviation is expressed in units of observation values, but standard distance is expressed in distance units, which are a function of the coordinate system or projection adopted.

The standard distance of a point distribution can be calculated by using the following equation:

$$SD = \sqrt{\frac{\sum_{i=1}^{n}(x_i - x_{mc})^2 + \sum_{i=1}^{n}(y_i - y_{mc})^2}{n}},$$

where (x_{mc}, y_{mc}) is the mean center of the point distribution.

Since points in a distribution may have attribute values that can be used as weights when calculating their mean center or even their median center, it is also possible to weight the points with specified attribute values when calculating the standard distance. For *weighted standard distance*, the following equation can be used:

$$SD = \sqrt{\frac{\sum_{i=1}^{n} f_i(x_i - x_{mc})^2 + \sum_{i=1}^{n} f_i(y_i - y_{mc})^2}{\sum_{i=1}^{n} f_i}},$$

where f_i is the weight for point, (x_i, y_i).

Using the 5 Ohio cities selected from the list of 125 largest U.S. cities, the standard distance is derived and the associated standard distance circle is presented in Figure 2.1. The steps for manually calculating the standard distance and the weighted standard distance are shown in Table 2.3.

TABLE 2.3 Standard Distance and Weighted Standard Distance of 5 Ohio Cities from the Largest 125 U.S. Cities

City Name	Longitude in Degrees x	Latitude in Degrees y	Population in 1990 p
Akron	−81.5215	41.0804	223,019
Cincinnati	−84.5060	39.1398	364,040
Cleveland	−81.6785	41.4797	505,616
Columbus	−82.9874	39.9889	632,910
Dayton	−84.1974	39.7791	182,044
$\sum$	$\sum x = -414.8908$	$\sum y = 201.4679$	$\sum p = 1,907,629$

$$\bar{x}_{mc} = \frac{-414.8908}{5} = -82.9782, \qquad \bar{y}_{mc} = \frac{201.4679}{5} = 40.2936$$

City Name	$x - x_{mc}$	$(x - x_{mc})^2$	$p(x - x_{mc})^2$
Akron	+1.4567	2.1220	473,246.3180
Cincinnati	−1.5278	2.3342	849,742.1680
Cleveland	+1.2997	1.6892	854,086.5472
Columbus	−0.0092	0.0001	63.2910
Dayton	−1.2192	1.4864	270,590.2016
$\sum$		7.6319	2,450,728.5260

City Name	$y - y_{mc}$	$(y - y_{mc})^2$	$p(y - y_{mc})^2$
Akron	+0.7868	0.6191	138,071.0629
Cincinnati	−1.1538	1.3313	484,646.4520
Cleveland	+1.1861	1.4068	711,300.5888
Columbus	−0.3047	0.0928	58,734.0480
Dayton	−0.5145	0.2647	48,187.0468
$\sum$		3.7147	1,440,939.1990

(continued)

TABLE 2.3 *Continued.*

Standard Distance

Step 1:
$$\sum(x - x_{mc})^2 = 7.6319$$
$$\sum(y - y_{mc})^2 = 3.7147$$

Step 2:
$$SD = \sqrt{\frac{\sum_{i=1}^{n}(x_i - x_{mc})^2 + \sum_{i=1}^{n}(y_i - y_{mc})^2}{n}}$$

$$= \sqrt{\frac{7.6319 + 3.7147}{5}}$$

$$= \sqrt{\frac{11.3466}{5}}$$

$$= \sqrt{2.2693}$$

$$= 1.5064$$

Weighted Standard Distance

Step 1:
$$\sum p(x - x_{mc})^2 = 2,450,728.526$$
$$\sum p(y - y_{mc})^2 = 1,440,939.199$$

Step 2:
$$SD_w = \sqrt{\frac{\sum_{i=1}^{n} f_i(x_i - x_{mc})^2 + \sum_{i=1}^{n} f_i(y_i - y_{mc})^2}{\sum_{i=1}^{n} f_i}}$$

$$= \sqrt{\frac{2,450,728.526 + 1,440,939.199}{1,907,629}}$$

$$= \sqrt{\frac{3,891,667.725}{1,907,629}}$$

$$= \sqrt{2.0401}$$

$$= 1.4283$$

ArcView Notes In terms of its application, standard distance is usually used as the radius to draw a circle around the mean center to give the extent of spatial spread of the point distribution it is based on. In the project file Ch2.apr, the added item under Spatial Analysis calculates the mean center before calculating the standard distance. It then uses the standard distance as the radius to draw a circle around the mean center. The script also provides the options for users to save the mean center with or withou the standard distance as point and polygon shapefiles for future use.

Different standard distance circles can be drawn for different types of events or incidences in the same area. The same types of events or incidences can also be drawn in different areas. All these can provide the basis for visual comparison of the extent of spatial spread among different types of events or different areas. Between two neighborhoods with the same number of houses, the neighborhood that has a longer standard distance is obviously spreading over more space geographically than the other neighborhood. Similarly, for all cities in a state, standard distances will be different if their calculation is weighted by different attributes, such as their population sizes.

While similar applications can be structured easily, it is important to understand that comparisons of standard distances between point distributions may or may not be meaningful. For instance, the standard distance of Japan's largest cities weighted by population counts is calculated as 3.27746 decimal degrees, while it is 8.84955 decimal degrees for Brazil's largest cities. If the two standard distances are used alone, they indicate that Brazil has a much more dispersed urban structure than Japan. However, given the very different sizes and territorial shapes of the two countries, the absolute standard distances may not reflect accurately the spatial patterns of how the largest cities spread in these countries.

To adjust for this possible bias, the standard distance may be scaled by the average distance between cities in each country or by the area of each country or region. Alternatively, the standard distances can be standardized or weighted by a factor that accounts for the territorial differences of the two countries. In general, standard distances can be scaled by a variable that is a function of the size of the study areas. In this example, the weighted standard distances are 0.2379 and 0.0272 for Japan and Brazil, respectively, when scaled by their areas, indicating that Japan's largest cities are in fact more dispersed than Brazil's cities.

2.3.2   Standard Deviational Ellipse

The standard distance circle is a very effective tool to show the spatial spread of a set of point locations. Quite often, however, the set of point locations may come from a particular geographic phenomenon that has a directional bias. For instance, accidents along a section of highway will not always form a circular shape represented by a standard distance circle. Instead, they will appear as a linear pattern dictated by the shape of that section of highway. Similarly, occurrences of algae on the surface of a lake will form patterns that are limited by the shape of the lake. Under these circumstances, the standard distance circle will not be able to reveal the directional bias of the process.

A logical extension of the standard distance circle is the *standard deviational ellipse*. It can capture the directional bias in a point distribution. There are three components in describing a standard deviational ellipse: the angle of rotation, the deviation along the major axis (the longer one), and the deviation along the minor axis (the shorter one). If the set of points exhibits certain directional bias, then there will be a direction with the maximum spread of the points. Perpendicular to this direction is the direction with the minimum spread of the points. The two axes can be thought of as the *x* and *y* axes in the Cartesian coordinate system but rotated to a particular angle corresponding to the geographic orientation of the point distribution. This angle of rotation is the angle between the north and the *y* axis rotated clockwise. Please note that the rotated *y* axis can be either the major or the minor axis. Figure 2.4 illustrates the terms related to the ellipse.

The steps in deriving the standard deviational ellipse are as follows:

1. Calculate the coordinates of the mean center, (x_{mc}, y_{mc}). This will be used as the origin in the transformed coordinate system.

2. For each point, p_i, in the distribution, transform its coordinate by

$$x'_i = x_i - x_{mc}$$

$$y'_i = y_i - y_{mc}.$$

After they have been transformed, all points center at the mean center.

3. Calculate the angle of rotation, θ, such that

$$\tan \theta = \frac{\left(\sum_{i=1}^{n} x'^2_i - \sum_{i=1}^{n} y'^2_i \right) + \sqrt{\left(\sum_{i=1}^{n} x'^2_i - \sum_{i=1}^{n} y'^2_i \right)^2 + 4 \left(\sum_{i=1}^{n} x' \sum_{i=1}^{n} y' \right)^2}}{2 \sum_{i-1}^{n} x'_i \sum_{i=1}^{n} y'_i}.$$

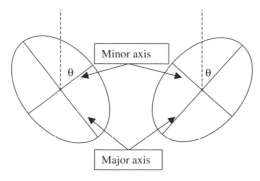

Figure 2.4   Elements defining a standard deviational ellipse.

$\tan\theta$ can be positive or negative. If the tangent is positive, it means that the rotated y axis is the longer or major axis and rotates clockwise from north. If the tangent is negative, it means that the major axis rotates counterclockwise from north. If the tangent is positive, we can simply take the inverse of tangent θ (arctan) to obtain θ for subsequent steps. If tangent is negative, taking the inverse of the tangent of a negative number will yield a negative angle (such as $-x$), i.e., rotating counterclockwise. But angle of rotation is defined as the angle rotating clockwise to the y axis, therefore, 90 degrees have to be added to the negative angle (i.e., $90 - x$) to derive θ. With θ from step 3, we can calculate the deviation along the x and y axes in the following manner:

$$\delta_x = \sqrt{\frac{\sum_{i=1}^{n} \left(x_i' \cos\theta - y_i' \sin\theta\right)^2}{n}}$$

and

$$\delta_y = \sqrt{\frac{\sum_{i=1}^{n} \left(x_i' \sin\theta - y_i' \cos\theta\right)^2}{n}}.$$

ArcView Notes The script for the standard deviational ellipse incorporated in the project file derives all three parameters necessary to fit a standard deviational ellipse to a set of points. While ArcView 3.1 and later versions support the ellipse object, ellipses slightly different in orientation cannot be easily detected visually. Instead, the script uses two perpendicular axes to show the orientation of the points and the spatial spread along the two directions.

2.4 APPLICATION EXAMPLES

We have discussed the concepts and background of a set of centrographic measures for analyzing point data. Although these measures are very useful in analyzing point data, they have not been used as widely as expected. We can still find many examples using these geostatistics in various research publications or applications. After each census in the United States, the Bureau of the Census calculates both the mean center and the median center of the entire U.S. population. Between censuses, the estimated mean center and median are also reported (for example, in the Census, 1996). By plotting the mean centers and median centers for each census year in the past century, it shows that the center of the U.S. population has been moving from the Northeast (Maryland and Delaware) to the West and the Southwest. Today the mean center is somewhere near St. Louis, Missouri.

Thapar et al. (1999) calculated mean population centers of the United States at two different geographical scales: census regions and states. By comparing the mean centers over different censuses, the results help us to depict the gross pattern of population movement at different spatial scales. Thapar et al. (1999) also reviewed several other studies using mean center as a descriptor tool for point data. In another example, Greene (1991) was concerned about the spread of economically disadvantaged groups over time. After deriving the standard distances of these population groups for different cities, Greene created circles based upon the standard distances to compare the location and geographical spread of these groups in several cities over time. In a completely different context, Levine et al. (1995) applied standard deviational ellipses to compare different types of automobile accidents in Hawaii. This study was done to decide if there was any directional bias among the types of accidents. The authors ultimately were able to provide a prescription for the local authority to deal with some "hot spots." By fitting an ellipse to a specific ethnic group, different ellipses are derived for different groups. These ellipses can be laid over each other to indicate the extent of their spatial correspondence. Using this idea, Wong (1999) recently derived a spatial index of segregation. The ellipse-based index was also used to study the spatial integration of different ethnic groups in China (Wong, 2000).

The articles reviewed above are not meant to be an exhaustive list. Readers can identify additional studies using these measures. To illustrate how these measures can be used in ArcView-accompanied Avenue scripts incorporated into the project file (CH2.APR) in this book, a point data set will be analyzed to show how these statistics can be use for real-world data.

2.4.1 Data

The point data set has been selected from the U.S. Environmental Protection Agency Toxic Release Inventory (EPA, 1999). The details of this database and related data quality issues are presented on EPA's website (*http://www.epa.gov*), so they will not be discussed here. Interested readers should carefully review all the data documentation and metadata before using this database in formal analysis and research. In brief, the TRI database is organized by state. Submitted to the EPA by each state government for the TRI database is the information about facilities that produce certain types of substances and chemicals that are known to be hazardous to the environment and the well-being of humans. The TRI database lists the facilities and their characteristics, such as chemicals released to the environment through various channels, together with their locations expressed in longitude and latitude.

We will focus on the TRI facilities in the state of Louisiana. Like most databases, the TRI database also has errors in positional accuracy. After sites with erroneous locational information are eliminated, there are 2,102 records in the Louisiana database. Among these 2,102 records, there are 234 types of chemicals. For the purpose of this exercise, only the longitude, latitude, and chemical information were extracted from the database to be imported into ArcView.

ArcView Notes The three columns of TRI data (longitude, latitude, and chemical name) have to be converted into **dBase IV** format so that they can be imported into ArcView. Many spreadsheet and database packages can perform this conversion. With a new project opened in ArcView, the dBase table can first be added to the project to create an ArcView Table document. Click the **Table** icon in the project window and then click the **Add** button to bring up the **Add Table** window. Navigate to the directory where the dBase file is stored. Double click the filename to bring it in as an ArcView Table document.

In creating a point theme of the TRI facilities, a new View document was first created. In this View document, the **Add Event Theme** function under the **View** menu item was used to geocode the TRI facilities by choosing the Table document that contains the TRI information as the geocode table. After the point theme was created by the geocode procedure, it was converted into a shapefile (`la_tri.shp`) using the **Convert to Shapefile** function item under the **Theme** menu. Please note that in the table shown in Figure 2.5, two additional columns were added to the original three columns of data. The first new column is the **Shape** field, which is listed as "point." This is a column that ArcView creates and adds to the data set in order to define the geographical object (point, polyline, or polygon) represented in that theme. The last column will be discussed later.

To obtain a list of all chemicals and the number of facilities releasing these chemicals, we can summarize the table shown in Figure 2.5. First, highlight (click) the `chem_name` field. Then go to the **Field** menu to choose the **Summarize** function. In the **Field** window of the dialog box, choose either **Lat** or **Long**. Then in the **Summarize-by** window, choose **Count**. Click the **Add** button so that a `Count_lat` or `Count_long` query can be added to the window on the right. If you want to designate the summary table to be saved in a particular location, make the changes in the **Save-As** window. Part of the summary table is shown in Figure 2.6. The table shown is sorted in descending order according to the numbers of facilities releasing any chemicals. From the table, it is clear that the chemicals released by most facilities are chlorine, ammonia, toluene and methanol.

2.4.2 Analyses: Central Tendency and Dispersion

Although more than 200 types of chemicals are reported in the Louisiana TRI database, it is obvious that some chemicals concentrate in specific locations rather than spreading evenly across the state. The set of statistics described in this chapter can help explore the spatial characteristics of this point data set. Conceptually, we could derive a mean center, a standard distance and the associated circle, and

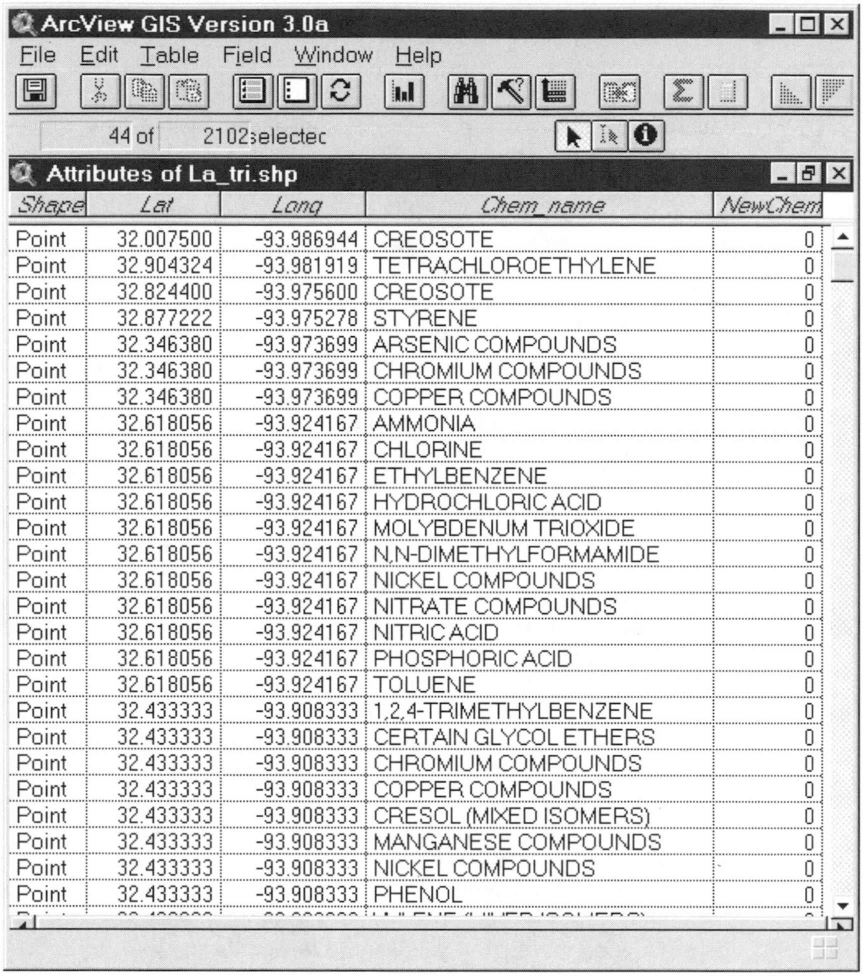

Shape	Lat	Long	Chem_name	NewChem
Point	32.007500	-93.986944	CREOSOTE	0
Point	32.904324	-93.981919	TETRACHLOROETHYLENE	0
Point	32.824400	-93.975600	CREOSOTE	0
Point	32.877222	-93.975278	STYRENE	0
Point	32.346380	-93.973699	ARSENIC COMPOUNDS	0
Point	32.346380	-93.973699	CHROMIUM COMPOUNDS	0
Point	32.346380	-93.973699	COPPER COMPOUNDS	0
Point	32.618056	-93.924167	AMMONIA	0
Point	32.618056	-93.924167	CHLORINE	0
Point	32.618056	-93.924167	ETHYLBENZENE	0
Point	32.618056	-93.924167	HYDROCHLORIC ACID	0
Point	32.618056	-93.924167	MOLYBDENUM TRIOXIDE	0
Point	32.618056	-93.924167	N,N-DIMETHYLFORMAMIDE	0
Point	32.618056	-93.924167	NICKEL COMPOUNDS	0
Point	32.618056	-93.924167	NITRATE COMPOUNDS	0
Point	32.618056	-93.924167	NITRIC ACID	0
Point	32.618056	-93.924167	PHOSPHORIC ACID	0
Point	32.618056	-93.924167	TOLUENE	0
Point	32.433333	-93.908333	1,2,4-TRIMETHYLBENZENE	0
Point	32.433333	-93.908333	CERTAIN GLYCOL ETHERS	0
Point	32.433333	-93.908333	CHROMIUM COMPOUNDS	0
Point	32.433333	-93.908333	COPPER COMPOUNDS	0
Point	32.433333	-93.908333	CRESOL (MIXED ISOMERS)	0
Point	32.433333	-93.908333	MANGANESE COMPOUNDS	0
Point	32.433333	-93.908333	NICKEL COMPOUNDS	0
Point	32.433333	-93.908333	PHENOL	0

Figure 2.5 Portion of the ArcView attribute table showing the TRI data of Louisiana.

a standard deviational ellipse for each chemical. To illustrate the utility of the geostatistics and associated concepts, we choose only two chemicals: copper and ethylene. Using the Avenue script built into the project file `CH2.APR`, a mean center was created for each chemical. Based upon each mean center, a standard distance was derived for each of the two chemicals. With the standard distances as radii, two standard distance circles were drawn. The results are shown in Figure 2.7.

In Figure 2.7, all the TRI sites are included, with those releasing copper and ethylene represented by different symbols. Also note that some symbols are on top of each other because those facilities released multiple chemicals. As shown in

Chem_name	Count	Count_Lat
CHLORINE	78	78.000000
AMMONIA	76	76.000000
TOLUENE	76	76.000000
METHANOL	75	75.000000
XYLENE (MIXED ISOMERS)	64	64.000000
PHOSPHORIC ACID	56	56.000000
HYDROCHLORIC ACID	54	54.000000
ZINC COMPOUNDS	52	52.000000
BENZENE	45	45.000000
ETHYLBENZENE	44	44.000000
ETHYLENE	44	44.000000
N-HEXANE	40	40.000000
SULFURIC ACID	38	38.000000
NAPHTHALENE	38	38.000000
METHYL ETHYL KETONE	36	36.000000
PROPYLENE	34	34.000000
NITRATE COMPOUNDS	34	34.000000
FORMALDEHYDE	32	32.000000
STYRENE	32	32.000000
CERTAIN GLYCOL ETHERS	30	30.000000
PHENOL	29	29.000000
ETHYLENE GLYCOL	28	28.000000
1,2,4-TRIMETHYLBENZENE	28	28.000000
NICKEL COMPOUNDS	27	27.000000
CYCLOHEXANE	24	24.000000
COPPER COMPOUNDS	23	23.000000

Figure 2.6   Portion of the table summarizing the TRI table by chemicals.

Figure 2.7, most TRIs releasing copper are found in the central and northwestern portions of the states, while most TRIs releasing ethylene are closer to the Gulf of Mexico. A comparison of the mean centers for the two sets of points shows that in general, TRIs releasing copper are mostly located in areas northwest of the TRIs releasing ethylene. But as indicated by the two standard distance circles (1.605 for copper and 1.159 for ethylene as their radii), the standard distance circle for copper TRIs is larger than the one for ethylene TRIs. In other words, TRIs releasing copper are geographically more dispersed in the northwestern part of the state than those releasing ethylene concentrating in the southeast.

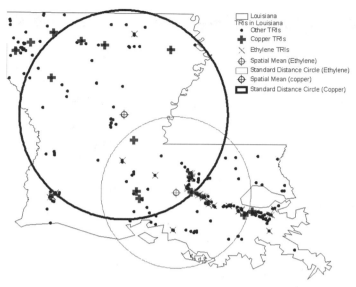

Figure 2.7   Spatial means and standard distances for TRIs in Louisiana releasing copper and ethylene.

In this example, because the two sets of points are from the same geographical region, we do not have to make any adjustment in the geographical scale when we compare the two standard distance circles.

ArcView Notes The summary table shown in Figure 2.6 indicates that 13 facilities released copper (not shown in the table) and 44 facilities released ethylene. These facilities are then reclassified into three groups by adding a new attribute, Newchem, to the attribute table. The three new categories are copper (1), ethylene (2), and others (0). Using the **Query Builder** button in ArcView, we can select facilities according to the values in the Newchem field. By selecting Newchem = 1 in the **Query Builder** window, we select all TRIs releasing copper. Choosing the menu item **Spatial Mean/Standard Distance** under the **Spatial Analysis** menu will invoke the script to derive the mean center and draw the standard distance circle. The script is designed in such a way that if no points or records are selected, all the points or records will be selected and entered into the calculation of the statistics. After the script is run, a mean center and a standard distance circle will be generated for copper TRI sites. The process can be repeated for ethylene TRI sites.

Note that the calculation of standard distance is based upon the map units defined under **View Properties**. If the map units are in degree-decimal, the resultant standard distance is also in degree-decimal. In general, for large study areas, the data should be projected instead of using latitude-longitude. For small study areas, the distortion due to the unprojected coordinates may be insignificant. Conceptually, we could also derive the median center for each of the two chemicals, but we feel that median center will not provide additional insights into the problem.

2.4.3   Analyses: Dispersion and Orientation

In addition to mean center and standard distance circles, we can analyze the two sets of points using the standard deviational ellipses. Figure 2.8 shows the two ellipses indicated by their axes. The angle of rotation (i.e., the angle from north clockwise to the axis) for the copper ellipse is 44.6 degrees, while for the ethylene ellipse the angle of rotation is only 6.39 degrees. Obviously, the ellipse for ethylene TRIs follows an east-west direction and is relatively small compared to the ellipse for copper TRIs, which is larger and follows a northwest-southeast direction. The difference in orientation of these two sets of points is clearly depicted by the two ellipses.

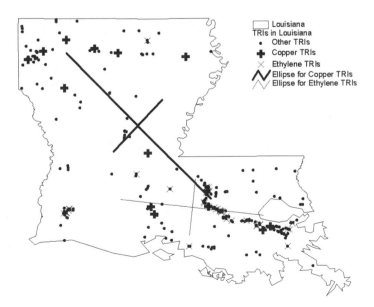

Figure 2.8   The standard deviational ellipses for TRIs releasing copper and ethylene.

> **ArcView Notes** The setup of the accompanying script to fit the deviational ellipse is similar to that of the other scripts for point descriptors. Users can use the **Query Builder** to select different sets of point to be fitted by the ellipses. This script, however, will produce additional parameters defining the ellipse. These parameters include $\tan\theta$, the lengths of the two axes, and the angle of rotation. As discussed in Section 2.3.2, the angle of rotation has to be adjusted if $\tan\theta$ is negative. The script will make this adjustment automatically when the angle of rotation is reported. In addition, if the script reports a negative $\tan\theta$, it means that the major axis is counterclockwise from the north.
>
> The existing setup of the project file `Ch2.apr` allows users to bring themes into this project for analysis using the built-in functions of the project file. To do this, start ArcView and then use the **Open Project** menu item in the **File** menu to open `Ch2.apr`. When calculating any of the geostatistics discussed in this chapter, use the **Add Theme** menu item in **View** menu to bring in geo-datasets and then choose appropriate menu items from the **Spatial Analysis** menu.

Even though centrographic measures are very useful in extracting spatial patterns and trends in point sets, there are pitfalls readers should be aware of. First, the statistics may be limited by the boundary effect. In the example here, TRI facilities may not follow state lines between Louisiana and its neighboring states, and the use of the state boundary here may or may not cut off the distribution of TRI facilities. The decision to use this boundary normally requires careful consideration, as the orientation of the boundary may force the point distribution to have spatial patterns that do not exist in reality.

No matter how carefully the analysis is performed, if the data are not properly sampled or properly processed, biased or inaccurate results will cause analyst to make incorrect decisions. This is particularly true in geo-referenced data. For point data, the analyst should explore whether or not the attribute values of any given point are affected by the neighboring points in some way. That is, the analyst must determine if the set of points contains any spatial pattern that is worthy of study.

In later chapters, we will discuss methods for detecting and measuring such spatial patterns. The methods discussed in this chapter are descriptive. They are used to describe and compare point sets as the first step in geographic analysis.

REFERENCES

Bureau of the Census. (1996). *Statistical Abstract of the United States 1995*. Washington, DC: U.S. Bureau of the Census.

Environmental Protection Agency (EPA). (1999). *http://www.epa.gov/opptintr/tri/*

Ebdon, D. (1988). *Statistics in Geography*. New York: Basil Blackwell.

Greene, R. (1991). Poverty concentration measures and the urban underclass. *Economic Geography*, 67(3):240–252.

Kellerman, A. (1981). *Centrographic Measures in Geography. Concepts and Techniques in Modern Geography (CATMOG) No. 32*. Norwich: Geo Book, University of East Angolia.

Levine, N., Kim, K. E., and Nitz, L. H. (1995). Spatial analysis of Honolulu motor vehicle crashes: I. Spatial patterns. *Accident Analysis and Prevention*, 27(5):675–685.

Monmonier, M. (1993). *Mapping it Out*. Chicago: University of Chicago Press.

Thapar, N., Wong, D., and Lee, J. (1999). The changing geography of population centroids in the United States between 1970 and 1990. *The Geographical Bulletin*, 41:45–56.

Wong, D. W. S. (1999). Geostatistics as measures of spatial segregation. *Urban Geography*, 20(7):635–647.

Wong, D. W. S. (2000). Ethnic integration and spatial segregation of the Chinese population. *Asian Ethnicity*, 1:53–72.

CHAPTER 3

PATTERN DETECTORS

The descriptive spatial statistics introduced in the previous chapter are useful in summarizing point distributions and in making comparisons between point distributions with similar attributes. But the use of these descriptive statistics is only the first step in geographic analysis. The next step is to analyze a point distribution to see if there is any recognizable pattern. This step requires additional tools, such as those to be discussed in this chapter.

Every point distribution is the result of some processes at a given point in time and space. To fully understand the various states and the dynamics of particular geographic phenomena, analysts need to be able to detect spatial patterns from the point distributions. This is because successful formulation or structuring of spatial processes often depends to a great extent on the ability to detect changes between point patterns at different times or changes between point patterns with similar characteristics. The recognition and measurement of patterns from point distributions is therefore a very important step in analyzing geographic information.

In considering how cities distribute over a region, one can easily find situations in which cities distribute unevenly over space. This is because the landscape, transportation network, natural resources, and possibly many other factors might have influenced the decision to choose different locations for the settlements to start with and the different growth rates these settlements might have had afterward.

At a global or continental scale, cities are often represented as points on a map. At a local scale, incidences of disease or crime in a city or incidences of fire in a forest may be plotted similarly. When studying point distributions such as these, analysts may try to relate them to particular patterns based on their experience or knowledge generated from previous studies, particularly studies that have developed theories and models. One example is to examine how cities distribute in a region, resembling the theoretical pattern (hexagonal pattern) of city distribution under the Central Place Theory. To do so, analysts would need tools that do more than summarize the statistical properties of point distributions.

The techniques that we discuss in this chapter, although having limitations, are useful in detecting spatial patterns in point distributions (Getis and Boots, 1988). We will first introduce Quadrat Analysis, which allows analysts to determine if a point distribution is similar to a random pattern. Next, the nearest neighbor analysis compares the average distance between nearest neighbors in a point distribution to that of a theoretical pattern (quite often a random pattern). Finally, the spatial autocorrelation coefficient measures how similar or dissimilar an attribute of neighboring points is.

3.1 SCALE, EXTENT, AND PROJECTION

The methods to be discussed in this chapter are used mainly to detect or measure spatial patterns for point distributions. It is necessary to pay special attention to three critical issues when using these methods.

First, we need to choose a proper geographic scale to work with when using points to represent some geographic objects. This is because geographic objects may be represented differently at different scales, depending on how they are treated. Whether to hold the scale constant in a study of certain geographic objects or to allow the scale to be changed is a rather important issue to consider when working with sets of points scattering over space.

As pointed out earlier, cities are often represented by points at a global or continental scale. The City of Cleveland, for example, appears to be only a point when a map shows it with other major cities in the United States. The same city, however, becomes a polygonal object that occupies an entire sheet of map when a larger-scale map shows the city with all its streets, rivers, and other details. Similarly, a river may be represented in a small-scale map as a linear feature, but it is an ecosystem if an analyst's focus is on its water, riverbeds, banks, and all of its biological aspects.

The second issue is the extent of geographic areas in the study. Analysts often need to determine to what extent the areas surrounding the geographic objects of interest are to be included in their analysis. Let's assume that we are working on a study that examines intercity activities including Akron, Cincinnati, Cleveland, Columbus, and Dayton in Ohio. When only the geographic extent of the state of Ohio is used, the five cities seem to scatter quite far apart from each other, as shown in Figure 3.1. However, these five cities would seem to be very closely clustered if we define the entire United States to be the study area with respect to them. To increase this difference further, the five cities essentially cluster nearly at one location if they are considered from the perspective of the entire world.

Figure 3.1 Comparative clusterness.

Delimiting the study area properly for a project is never easy. There are cases in which political boundaries are appropriate choices, but there are also cases in which they are not. While no single, simple solution can be offered for all cases, analysts are urged to consider this issue carefully.

The last issue is the projection used in maps when displaying the distribution of geographic objects. Four possible distortions may be caused by different projections: *area*, *shape*, *direction* and *distance*. Among the more than 20 different projections that can be applied in mapping, there is no projection that can perfectly transform geographic locations from their locations on the globe (a three-dimensional space) to a map (a two-dimensional plane). Therefore, we will need to be sensitive to the needs of the project or study at hand. Careful consideration of the purposes, scales, and accuracy of data is needed for a successful study.

In detecting point patterns, it is important to consider the impact of different projections. This is because both area and distance are used intensively in the analysis of point patterns. In Quadrat Analysis, the size of the study area affects the density of points. In the nearest neighbor analysis and the spatial autocorrelation coefficient, distances between points play a critical role in the calculation of these two indices.

Not surprisingly, the larger the study area, the more significant the impact of different projections will be. If the study area is small, such as a residential neighborhood or a small village/city the different areal and distance measurements by different projections may not matter. However, studies that encompass the entire

United States or the world have to pay special attention to selecting proper projections to work with.

ArcView Notes Two types of projections are most popular when working with data concerning the United States. They are known as *State Plane* projections and *Universal Mercator Projections* (*UTM*). When working with projects at local scales, e.g., one or several counties, analysts often use the State Plane projection to display their data in **View** documents or in maps created in **Layout** documents. For projects dealing with an entire state, UTM is often a better choice. In projects dealing with data at a global or continental scale, the choice of projections will depend on where the areas are located and the purpose of the analysis.

As an example, one can see that the shape of the mainland United States changes between projections in Figure 3.2a. Consequently, these different shapes of the United States will affect the measurements of area and distance between places. To use proper projections in a study, factors such as areal extent, the purpose of the study, and the location of the study area need to be assessed carefully.

Compared to the entire United States, the changes caused by changing projections in the case of the State of Ohio are much less significant. The projections showed in Figure 3.2b display few changes compared to the extent of the area of concern.

Although ArcView can be specified to manage projected data, we recommend adding unprojected data (i.e., the data are in the latitude-longitude geo-referencing system) as themes to View documents if possible. This is because ArcView provides more flexibility in allowing its users to change a View document's projection if the themes are added as unprojected data. To do this,

1. access the menu item **View** and the **Properties** from the drop-down menu list;
2. specify the proper data unit and map unit based on your data;
3. use the **Projection** button to access the dialog box for changing projections;

Equal-Area Projection   Un-Projected   Equal-Distance Projection

Figure 3.2a Examples of distortions caused by changing projections.

Un-Projected   State Plane Projection   UTM Projection

Figure 3.2b   Examples of distortions caused by changing projections.

4. choose the projection category based on the geographic extent of your data and then, as required, choose the projection zone for your study area.

3.2   QUADRAT ANALYSIS

The first method for detecting spatial patterns from a point distribution is *Quadrat Analysis*. This method evaluates a point distribution by examining how its density changes over space. The density, as measured by Quadrat Analysis, is then compared with usually, but not necessarily a theoretically constructed random pattern to see if the point distribution in question is more clustered or more dispersed than the random pattern.

The concept and procedures of Quadrat Analysis are relatively straightforward. First, the study area is overlain with a regular square grid, and the number of points falling in each of the squares are counted. Some squares will contain none of the points, but other squares will contain one, two, or more points. With all squares counted, a frequency distribution of the number squares with given number of points can be constructed. Quadrat Analysis compares this frequency distribution with that of a known pattern, such as a theoretically random pattern.

The squares are referred to as *quadrats*, but quadrats do not always need to be squares. Analysts can use other geometric forms, such as circles or hexagons, as appropriate for the geographic phenomenon being studied. The selection among various forms of quadrats should be based on previous successful experience or the characteristics of the phenomenon in question. In addition, within each analysis, the shape and size of the quadrats have to be constant.

In considering an extremely clustered point pattern, one would expect all or most of the points to fall inside one or a very few squares. On the other hand, in an extremely dispersed pattern, sometimes referred to as a *uniform* pattern, one would expect all squares to contain relatively similar numbers of points. As a

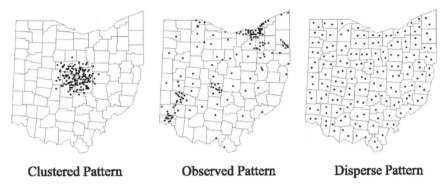

Clustered Pattern	Observed Pattern	Disperse Pattern

Figure 3.3   Ohio cities with hypothetical clustered and cispersed patterns.

result, analysts can determine if the point distribution under study is closer to a clustered, random, or dispersed pattern. As examples of clustered and dispersed patterns, Figure 3.3 shows the 164 cities in the state of Ohio, along with hypothetical clustered and dispersed patterns with the same number of points.

Overlaying the study area with a regular grid partitions the study area in a systematic manner to avoid over- or undersampling of the points anywhere. Since Quadrat Analysis evaluates changes in density over space, it is important to keep the sampling interval uniform across the study area. There is, however, another way to achieve the same effect. This involves randomly placing quadrats of a fixed size over the study area (Figure 3.4). Statistically, Quadrat Analysis will achieve a fair evaluation of the density across the study area if it applies a large enough number of randomly generated quadrats.

ArcView   In the accompanying project file, `Ch3.apr`, three types of quadrats
Notes   are used. Users can choose to use squares, hexagons, or circles. In addition, users can choose between placing a regular grid over the study area or throwing randomly generated quadrats over it.

Similar to `Ch1.apr` and `Ch2.apr`, `Ch3.apr` can be found on the companion website to this book. Use ArcView to open this project file before adding data themes to the View document.

The last issue that needs careful consideration when applying Quadrat Analysis is the size of quadrats. According to the Greig-Smith experiment (Greig-Smith, 1952) and the subsequent discussion by Taylor (1977, pp. 146–147) and Griffith and Amrhein (1991, p. 131), an optimal quadrat size can be calculated as follows:

$$\text{Quadrat size} = \frac{2 \cdot A}{n},$$

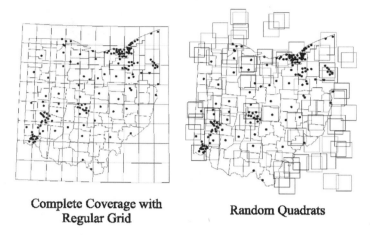

Complete Coverage with
Regular Grid

Random Quadrats

Figure 3.4   Quadrat analysis: complete coverage by grids and random quadrats.

where A is the area of the study area and n is the number of points in the distribution. This suggests that an appropriate square size has a width of $\sqrt{2 \cdot A/n}$.

Once the quadrat size for a point distribution is determined, Quadrat Analysis can proceed to establish the frequency distribution of the number of points in each quadrat, using either complete coverage (with a regular square grid or a hexagon grid) or randomly generated quadrats. This frequency distribution needs to be compared to a frequency distribution describing a random point pattern.

A statistical test known as the *Kolmogorov-Simirnov test* (or simply the *K-S test*), can be used to test the difference statistically between an observed frequency distribution and a theoretical frequency distribution. This test is a simple, straightforward test, both conceptually and computationally.

As an example, let's take the 164 Ohio cities and use 80 squares to construct the frequency distribution for Quadrat Analysis. The frequency distribution of the cities falling into squares is listed in Table 3.1. In this table, the left-hand column lists the number of cities in each square. The second column shows that there are 36 squares with no city at all, 17 squares with only one city, 10 squares with two cities, and so on. For an example of a uniform/dispersed pattern, the third column lists frequencies that are made up to approximate an even distribution of cities across all squares. The right-hand column indicates that all cities are located within one square.

By observing the differences among the three frequency distribution columns, it is clear that the observed pattern is more clustered than the dispersed pattern. But it is not as clustered as that of the right-hand column. While the differences among the columns can be visually estimated, we need some way of measuring the difference quantitatively. At this stage of our analysis, we can apply the K-S test.

The K-S test allows us to test a pair of frequency distributions at a time. Let's take the observed pattern and the dispersed pattern to start with. First, we assume

TABLE 3.1 Frequency Distribution: 164 Ohio Cities with Three Hypothetical Frequency Distributions

Number of Cities in Each Square	Observed Pattern	Uniform/Dispersed	Clustered
0	36	0	79
1	17	26	0
2	10	26	0
3	3	26	0
4	2	2	0
5	2	0	0
6	1	0	0
7	1	0	0
8	1	0	0
9	1	0	0
10	1	0	0
11	1	0	0
12	1	0	0
13	1	0	0
14	1	0	0
28	1	0	0
164	0	0	1

that the two frequency distributions are similar enough that we cannot detect differences between them that are statistically significant. This concept is slightly confusing for those who have not had much experience with statistical analysis, but it is simply making an assumption that allows a small degree of difference to be acceptable. If the difference between two frequency distributions is indeed very small, then the difference might have happened simply by chance. The larger the difference, the less likely that it occurred by chance.

The test is as follows:

1. Assume that there is no statistically significant difference between two frequency distributions.
2. Decide on a level of statistical significance—for example, allowing only 5 out of 100 times ($\alpha = 0.05$).
3. Convert all frequencies to cumulative proportions in both distributions.
4. Calculate the D statistic for the K-S test:

$$D = \max |O_i - E_i|,$$

where O_i and E_i are cumulative proportions of the ith category in the two distributions. The max $||$ term indicates that we do not care which one is larger than the other; we are concerned only with their difference. D is then the maximum of the absolute differences among all pairwise comparisons.

5. Calculate a critical value as the basis for comparison:

$$D_{\alpha=0.05} = \frac{1.36}{\sqrt{m}},$$

where m is the number of quadrats (or observations).
  In a 2-sample case,

$$D_{\alpha=0.05} = 1.36 \sqrt{\frac{m_1 + m_2}{m_1 m_2}}$$

where m_1 and m_2 are numbers of quadrats in the 2 groups.

6. If the calculated D is greater than the critical value of $D_{\alpha=0.05}$, we will conclude that the two distributions are significantly different in a statistical sense.

Taking the example of the 164 Ohio cities, Table 3.2 lists the frequencies and their converted proportions. The right-hand column lists the absolute differences between the two columns of cumulative proportions. The largest absolute difference, in this case, is 0.45. Therefore, $D = \max |O_i - E_i| = 0.45$.

Because this is a 2-sample case, the critical value $D(\alpha = 0.05)$ can be calculated as follows:

$$D(\alpha = 0.05) = 1.36 \sqrt{\frac{80 + 80}{80 * 80}} = 0.215.$$

TABLE 3.2 D Statistics for K-S Test

Number of Cities in Each Square	Observed Pattern	Cumulative Observed Proportions	Dispersed Pattern	Cumulative Pattern Proportions	Absolute Difference
0	36	0.45	0	0	0.45
1	17	0.66	26	0.325	0.34
2	10	0.79	26	0.65	0.14
3	3	0.83	26	0.975	0.15
4	2	0.85	2	1	0.15
5	2	0.88	0	1	0.13
6	1	0.89	0	1	0.11
7	1	0.90	0	1	0.10
8	1	0.91	0	1	0.09
9	1	0.93	0	1	0.08
10	1	0.94	0	1	0.06
11	1	0.95	0	1	0.05
12	1	0.96	0	1	0.04
13	1	0.98	0	1	0.03
14	1	0.99	0	1	0.01
28	1	1	0	1	0

The critical value of 0.215 is apparently far smaller than 0.45, indicating that the difference between two frequency distributions is statistically significant at the $\alpha = 0.05$ level. With this, we can easily reject our initial hypothesis that there is no significant difference between a dispersed pattern of 164 points and the distribution formed by the 164 Ohio cities. In other words, the 164 Ohio cities do not distribute in a dispersed manner.

In the above example, we examined the difference between an observed point pattern and a dispersed pattern. However, it is more common to compare an observed point pattern to a point pattern generated by a random process. A well-documented process for generating a random point pattern is the Poisson process. The Poisson random process is appropriate to generate count data or frequency distributions. Quadrat Analysis often compares an observed point pattern frequency distribution to a frequency distribution generated by the Poisson random process.

A Poisson distribution is strongly determined by the average number of occurrences, λ. In the context of Quadrat Analysis, λ is defined as the average number of points in a quadrat. Assume that we have m quadrats and n points in the entire study area, $\lambda = n/m$, the average number of points in a quadrat. Let x be the number of points in a quadrat. Using the Poisson distribution, the probability of having x points in a quadrat is defined as

$$p(x) = \frac{e^{-\lambda}\lambda^x}{x!}$$

where e is the Euler's 2.71828 constant and $x!$ is the factorial of x, which can be defined as $x(x-1)(x-2)\ldots(1)$ and $0!$, by definition, is 1. Using the probabilities for various values of x based upon the Poission distribution, we can generate a frequency distribution in the same format as those shown in Table 3.1 but for a random point distribution.

Generating a probability value from a Poisson distribution is rather simple. But if a set of probabilities for a range of x values is required, it becomes quite tedious, as the factorial and the e function have to be applied every time. Fortunately, there is a shortcut that can be used to generate a set of probabilities based upon the Poisson distribution. We know that if $x = 0$, the Poisson probability is reduced to

$$p(0) = e^{-\lambda}.$$

Other probabilities can be derived based upon $p(0)$. In general,

$$p(x) = p(x - 1) * \frac{\lambda}{x}.$$

If x is 1, then $p(x - 1) = p(0)$. Using this shortcut formula, it would be efficient to derive an entire set of Poisson probabilities.

In the Ohio example, there are 164 points (cities) and 80 quadrats were used in the previous example. Therefore, $\lambda = 164/80 = 2.05$. Using this λ value, a set of

TABLE 3.3   Comparing the 164 Ohio Cities to a Random Pattern Generated by a Poisson Process

Number of Cities in Each Square	Observed Pattern	Observed Probability	Cumulative Observed Probability	Poisson Probability	Cumulative Poisson Probability	Difference
0	36	0.45	0.4500	0.1287	0.1287	0.3213
1	17	0.2125	0.5514	0.2639	0.3926	0.2699
2	10	0.1125	0.7875	0.2705	0.6631	0.1244
3	3	0.0375	0.8250	0.1848	0.8480	0.0230
4	2	0.025	0.8500	0.0947	0.9427	0.0927
5	2	0.025	0.8750	0.0388	0.9816	0.1066
6	1	0.0125	0.8875	0.0133	0.9948	0.1073
7	1	0.0125	0.9000	0.0039	0.9987	0.0987
8	1	0.0125	0.9125	0.0010	0.9997	0.0872
9	1	0.0125	0.9241	0.0002	0.9999	0.0759
10	1	0.0125	0.9375	0.0000	1	0.0625
11	1	0.0125	0.9500	0.0000	1	0.0500
12	1	0.0125	0.9625	0.0000	1	0.0375
13	1	0.0125	0.9750	0.0000	1	0.0250
14	1	0.0125	0.9875	0.0000	1	0.0125
28	1	0.0125	1	0.0000	1	0.0000

Poisson probabilities can be derived using the shortcut formula. Table 3.3 shows the derivations of the probabilities. The first two columns in the table are identical to those in Table 3.1. The third and fourth columns are the observed probabilities and the cumulative probabilities based upon the observed pattern. In the fifth column, a Poisson distribution was generated based upon $\lambda = 2.05$. This probability distribution indicates the probability that a quadrat may receive different numbers of points. The cumulative probabilities of the Poisson distribution are also derived in the sixth column. The last column reports the absolute differences between the two sets of cumulative probabilities. The largest of these differences is the K-S D statistic, which is 0.3213, much greater than the critical value of 0.1520 using the 0.05 level of significance.

If the observed pattern is compared to a random pattern generated by a Poisson process, we can exploit a statistical property of the Poisson distribution to test the difference between the observed pattern and the random pattern in addition to the K-S statistic. This additional test is based upon the variance and mean statistics of a Poisson distribution.

We know that λ is the mean of a Poisson distribution. A very interesting and useful property of a Poisson distribution is that the variance is also λ. In other words, if a distribution potentially is generated by a random process like the Poisson process, the distribution should have the same mean and variance. If the mean and variance form a ratio, the variance-mean ratio, the ratio should be very close to 1. Therefore, given an observed point pattern and the frequency distribution of points by quadrats, we can compare the observed variance-mean ratio to 1 to see if they are significantly different. Given the information similar to that in Table 3.1, the mean of the observed distribution is $\lambda = 2.05$. The variance is basically a variance for grouped data. The mean λ has to be compared with the number of points in each quadrat. The difference is then squared and multiplied by the quadrat frequencies. The sum of these products divided by the number of quadrats produces the observed variance. Using the observed variance and λ, we can form a variance-mean ratio. This ratio is then compared to 1, and the difference has to be standardized by the standard error in order to determine if the standardized score of the difference is larger than the critical value (quite often 1.96 at the 0.05 level of significance).

The K-S test and the variance-mean ratio may yield inconsistent results. But because the K-S test is based upon weak-ordered data (Taylor, 1977) while variance-mean ratio is based upon an interval scale, the variance-mean ratio tends to be a stronger test. However, variance-mean ratio test can be used only if a Poisson process is expected.

Quadrat Analysis is useful in comparing an observed point pattern with a random pattern. Theoretically, we can compare the observed pattern with any pattern of known characteristics. For instance, after we compare the observed pattern with a random pattern and the result indicates that they are significantly different, the next logical step is to test if the observed pattern is similar to a clustering pattern or a dispersed pattern. Quite often, through visual inspection, the analyst can hypothesize what pattern the observed pattern resembles. Using other statistical distributions, such as the negative gamma or the negative binomial, we can generate point patterns with specific distribution properties. These patterns can then be compared with the observed pattern to see if they are different. Quadrat Analysis, however, suffers from certain limitations. The analysis captures information on points within the quadrats, but no information on points between quadrats is used in the analysis. As a result, Quadrat Analysis may be insufficient to distinguish between certain point patterns. Figure 3.5 is an example.

In Figures 3.5a and 3.5b, both spatial configurations have eight points with four quadrats. Visually, the two point patterns are different. Figure 3.5a is a more dispersed pattern, while Figure 3.5b is definitely a cluster pattern. Using quadrat

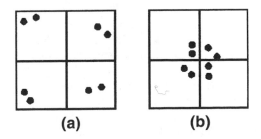

(a)   **(b)**

Figure 3.5   Local clusters with regional dispersion.

analysis, however, the two patterns yield the same result. In order to distinguish patterns depicted in Figure 3.5, we have to use Nearest Neighbor Analysis.

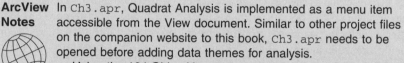

ArcView
Notes

In Ch3.apr, Quadrat Analysis is implemented as a menu item accessible from the View document. Similar to other project files on the companion website to this book, Ch3.apr needs to be opened before adding data themes for analysis.

Using the 164 Ohio cities as an example, Ch3.apr can be used to determine if that distribution is more clustered or more dispersed:

1. Download Ch3.apr from the companion website, and open it in ArcView.

2. In the View document, add two themes for Chapter 3 obtained from the website. These two themes are named ohcities.shp and ohcounty.shp. After they are added, change their display order, if necessary, so that the ohcities.shp is at the top of the Table of Contents in the View document.

3. Access the **View/Properties** menu item from the View document's menu. In the **View Properties** window, click the **Projection** button. In the **Projection Properties** window, change the projection **Category** to UTM-1983 and the projection **Type** to Zone 17. Click **OK** to return to the **View Properties** window. Click **OK** again to return to the View document.

4. There are 88 counties in the state of Ohio. Let's use approximately this many quadrats in our analysis since these counties are fairly similar in size. To start Quadrat Analysis, choose **Quadrat Analysis** from the drop-down menu of **Point Patterns**.

5. In the **Coverage Choice** window, choose **complete** and then click the **OK** button. Notice that the other choice is **random**, which is used when Quadrat Analysis is carried out using randomly generated quadrats.

6. In the **Shape Choice** window, choose **square** and then click the **OK** button. Similar to the previous step, two other options available. They are **hexagons** and **circles** (only for random coverage) in the drop-down list.

7. In the next **Input** window, we see that the number of points is 164. Enter **81** as the preferred number of quadrats.

8. Click the **OK** button for the next three windows as ArcView informs you of the length of the side of the quadrat (47,690.7

meters in our case), 9 vertical steps, and 9 horizontal steps. This is because the 81 quadrats form a 9 by 9 grid.

9. In the **Theme source FileName window**, navigate to your working directory and save the quadrat arrangement.

10. The first statistic is the **Lambda**, which is the average number of points per quadrat (in our case, 2.02469).

11. Next is the frequency, which is followed by a measurement of variance and then the **variance-mean ratio**.

12. In the **K-S D Statistic** window, we have, in our case, 0.361868.

13. For the level of **Alpha** (statistical significance), choose 0.05 and then click the **OK** button.

14. The calculated critical value is given as 0.1511 in our case. When this value is compared to 0.361868, the difference is statistically significant.

The resulting **View** document can be seen in Figure 3.6. The procedures we just went through placed a 9 by 9 grid over the study area.

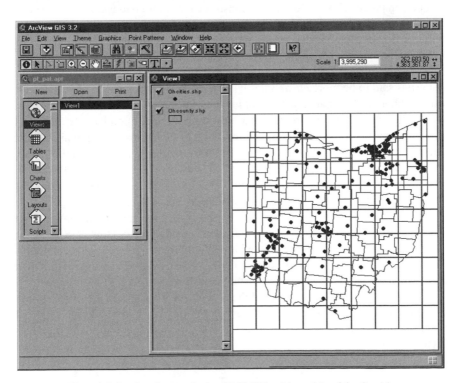

Figure 3.6 Quadrat analysis of 164 Ohio cities with a 9 by 9 grid.

As an alternative, it is possible to run the same project file for random quadrats. To do so, choose **random** in the **Coverage Choice** window. When asked for the number of quadrats preferred, give **81**. For the width of the square, give 25 (miles) since most counties are about this size.

Finally, the number of quadrats may be increased to ensure that the resulting frequency distribution in Quadrat Analysis approaches a normal frequency distribution. There is no number that is optimal in all cases, but the number of quadrats should be chosen based on the size of the study area, density of points, and computational time and resources.

3.3 NEAREST NEIGHBOR ANALYSIS

Quadrat Analysis tests a point distribution with the *points per area* (*density*) concept. The method to be discussed in this section, the *Nearest Neighbor Analysis*, uses the opposite concept of *area per point* (*spacing*). Quadrat Analysis examines how densities in a point distribution change over space so that the point pattern can be compared with a theoretically constructed random pattern. For the Nearest Neighbor Analysis, the test is based on comparing the observed average distances between nearest neighboring points and those of a known pattern. If the observed average distance is greater than that of a random pattern, we can say that the observed point pattern is more *dispersed* than a random pattern. Similarly, a point pattern is said to be more *clustered* if its observed average distance between nearest neighbors is less than that of a *random* pattern.

In a homogeneous region, the most uniform pattern formed by a set of points occurs when this region is partitioned into a set of hexagons of identical size and each hexagon has a point at its center (i.e., a triangular lattice). With this setup, the distance between points will be $1.075\sqrt{A/n}$, where A is the area of the region of concern and n is the number of points. This provides a good starting point for us to understand how the Nearest Neighbor Analysis works.

In the real world, we rarely see geographic objects distributing in an organized manner such as being partitioned by hexagons of equal size. However, we often see geographic objects, such as population settlements, animal/plant communities, or others distribute in a more irregular fashion. To test if any distribution had any recognizable patterns, let's use a statistic call R, named for randomness.

The R statistic, sometimes called the R *scale*, is the ratio of the observed average distance between nearest neighbors of a point distribution and the expected distance of the average nearest neighbor of the region of concern. It can be calculated as follows:

$$R = \frac{r_{\text{obs}}}{r_{\text{exp}}},$$

where r_{obs} is the observed average distance between nearest neighbors and r_{exp} is the expected average distance between nearest neighbors as determined by the theoretical pattern being tested.

To measure the observed average distance between nearest neighbors, we can calculate the distance between each point and all of its neighbor points. The shortest distance among these neighbors will be associated with the nearest point. When this process is repeated for all points, a table such as Table 3.4 can be calculated. The points in Table 3.4 are based on the distribution shown in Figure 3.7. For this set of 17 cities, the observed average distance between pairs of nearest neighbors is $r_{obs} = 6.35$ miles.

For the theoretical random pattern, let's use the following equation to calculate the expected average distance between nearest neighbors:

$$r_{exp} = \frac{1}{2\sqrt{n/A}},$$

where the n is the number of points in the distribution and the A is the area of the space of concern. In our example, the area of the five counties is 3,728 square miles. Therefore, the expected average distance is

$$r_{exp} = \frac{1}{2\sqrt{17/3728}} = 7.40.$$

TABLE 3.4   Observed Distances Between Nearest Neighbor Cities in the Five-County Area of Northeastern Ohio

City Name	Nearest City	Nearest Distance
Akron	Tallmadge	5.37
Alliance	North Canton	14.96
Barberton	Norton	2.39
Brunswick	Medina	7.93
Canton	North Canton	4.40
Cuyahoga Falls	Stow	4.52
Kent	Stow	4.36
Massillon	Canton	7.90
Medina	Brunswick	7.93
North Canton	Canton	4.40
Norton	Barberton	2.39
Portage Lakes	Barberton	4.00
Ravenna	Kent	6.32
Stow	Cuyahoga Falls	4.52
Tallmadge	Kent	4.38
Wadsworth	Norton	4.52
Wooster	Wadsworth	17.64
Average nearest distance		6.35 (miles)

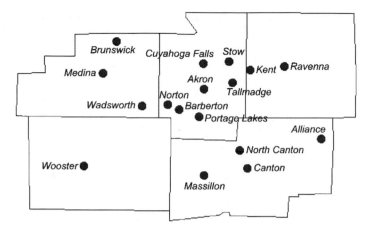

Figure 3.7 Cities in the five-county area of northeastern Ohio region.

With both distances calculated, we can now compute the R statistic:

$$R = \frac{r_{obs}}{r_{exp}} = \frac{6.35}{7.40} = 0.8581.$$

With this R scale, we know that the point pattern formed by the 17 cities is more clustered than a random pattern.

Now we know how the R statistic is calculated and how to determine if a pattern is more clustered or more dispersed than a random pattern. Many conclusions can be drawn from this calculation with regard to how the 17 cities relate to each other. But we are still not sure to what degree this pattern is more clustered than a random pattern. Is it much more clustered or just slightly more clustered? To appreciate the implications of various values of the R statistic, Figure 3.8 shows a series of hypothetical distributions and their associated R values.

Figure 3.8 shows that the more clustered patterns are associated with smaller R values ($r_{obs} < r_{exp}$) while the more dispersed patterns are associated with larger

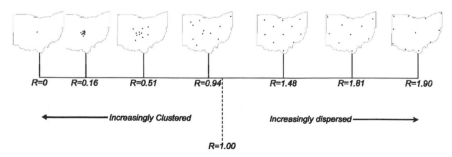

Figure 3.8 The scale of R statistics.

R values ($r_{\text{obs}} > r_{\text{exp}}$). Figure 3.8 is useful for establishing a general concept of how R values relate to various patterns. It is, however, not sufficient to measure quantitatively the difference between an observed pattern and a random pattern.

The R scale ranges from $R = 0$ (completely clustered) to $R = 1$ (random) to $R = 2.149$ (completely dispersed). When $R = 0$, all distances between points are zero, indicating that all points are found at the same location. When $R = 1$, $r_{\text{obs}} = r_{\text{exp}}$; the pattern being tested should therefore be a random pattern. When R approximates values or 2 or more, the pattern displays various degrees of dispersion.

When using Nearest Neighbor Analysis, one way to measure the extent to which the observed average distance differs from the expected average distance is to compare this difference with its *standard error* (SE_r). The standard error describes the likelihood that any differences occur purely by chance. If the calculated difference is relatively small when compared to its standard error, we say that this difference is not statistically significant. By contrast, when we have a difference that is relatively large with respect to its standard error, we claim that the difference is statistically significant; that is, it does not occur by chance.

The concept of standard error is rooted in classical statistical theories. In a normal distribution, there is about a 68% chance that some differences between one negative standard error and one positive standard error will occur by chance when in fact there should not be any difference between two populations being compared. Described in equation, this means that:

$$\text{Probability}(< 68\%) = (-1SE_r, +1SE_r).$$

Following this, we can define a calculated difference to be statistically significant only when it is smaller than $-1SE_r$ or greater than $+1SEE_r$. Or, if we want to be more rigid, we would call a difference statistically significant only if it is smaller than $-1.96SEE_r$ or greater than $1.96SEE_r$. This is because the probability of having a difference of that magnitude is 5 out of 100 times or less:

$$\text{Probability}(< 95\%) = (-1.96SE_r, +1.96SE_r).$$

To calculate the standard error for the observed distances, we can use the following equation:

$$SE_r = \frac{0.26136}{\sqrt{n^2/A}},$$

where n and A are as defined previously. With this standard error, we can now see how the difference is compared with it by calculating a standardized Z score:

$$Z_R = \frac{r_{\text{obs}} - r_{\text{exp}}}{SE_r}.$$

As mentioned earlier, if $Z_R > 1.96$ or $Z_R < -1.96$, we conclude that the calculated difference between the observed pattern and the random pattern is statistically significant. Alternatively, if $-1.96 < Z_R < 1.96$, we conclude that the observed pattern, although it may look somewhat clustered or somewhat dispersed visually, is not significantly different from a random pattern.

In our example of 17 Ohio cities, we have

$$SE_r = \frac{0.26136}{\sqrt{17^2/3728}} = 0.9387.$$

And the Z_R score is

$$Z_R = \frac{6.35 - 7.4}{0.9387} = \frac{-1.05}{0.9387} = -1.12,$$

meaning that it is not statistically different from a random pattern.

As demonstrated above, a point pattern may seem clustered or dispersed by visual inspection or even described by calculating its R value. However, we will not be able to reach a conclusion unless a statistical test confirms or rejects this conclusion. In other words, the calculated R value should be confirmed by Z_R scoring to ensure its statistical significance.

Please note that in the Ohio example, Z_R is negative, indicating that the nearest neighbor distance of the observed pattern is smaller than expected but insignificant. The sign of the z-score, however, indicates that the observed pattern has a clustering tendency. In other words, if the z-score indicates that the difference between the observed and expected nearest neighbor distances is statistically significant, the sign of the statistic can show if the observed pattern is probably clustered or dispersed. Following the logic of hypothesis testing, we can conduct a one-tailed test to see if the z-score is really negative (smaller than -1.645 at the 0.05 significance level) or really positive (greater than 1.645). These tests ultimately can provide a conclusion if the observed pattern is significantly different from a clustering pattern or a dispersed pattern.

With its ability to detect patterns in a point distribution, Nearest Neighbor Analysis has been extended to accommodate second, third, and higher order neighborhood definitions. When two points are not immediate nearest neighbors but are the second nearest neighbors, the way distances between them are computed will need to be adjusted accordingly. The extension is straightforward in concept and has been used on special occasions when this relationship is important.

For instance, using the first order nearest neighbor statistic, we cannot distinguish the two point patterns shown in Figure 3.5a and Figure 3.5b because the nearest neighbors of each point in both patterns are very close. But if the second order nearest neighbors are used in the analysis, the result will show that Figure 3.5a has a dispersed pattern because the second nearest neighbors are all far away in other quadrats. On the other hand, the result for Figure 3.5b will indicate a clustering pattern because all second nearest neighbors are still quite close. By combining the results of the first order and second order nearest neighbor analyses, we can conclude that Figure 3.5a has a local clustering but regional dispersed pattern, while Figure 3.5b has a clustering pattern on both local and regional scales. To a large extent, using different orders of nearest neighbor statistics can detect spatially heterogeneous processes at different spatial scales.

Finally, it should be noted that Nearest Neighbor Analysis has certain problems. When it is used to examine a point distribution, the results are highly sensitive to the geographic scale and the delineation of the study area. A set of cities may be considered very disperse if they are examined at a local scale. These same cities may seem extremely clustered if they are viewed at a continental or global scale. The 17 Ohio cities may seem quite dispersed if we use only the five counties as the study area. However, they are considered to be very clustered if they are plotted on a map that shows the entire United States.

Understanding the limitations of Nearest Neighbor Analysis, we should always be careful to choose an appropriate geographic scale to properly display the geographic objects we study. Furthermore, the delineation of study areas should be justified by meaningful criteria. In many cases political boundaries may make sense, but other situations may require boundaries defined by natural barriers such as coastlines, rivers, or mountain ranges.

ArcView Notes Nearest Neighbor Analysis is implemented in the ArcView project file `Ch3.apr` found on the companion website to this book. This project file has a customized menu item called **Point Patterns**. From **Point Patterns**, a drop-down menu provides access to all tools for detecting point patterns, as discussed in this chapter. To use Nearest Neighbor Analysis, bring in data layers by Add Theme and then choose **Point Patterns/Ordered Neighbor Statistics**.

As an example, we will use the 164 cities in Ohio as our data set to describe the procedures needed for calculating the R statistic and its standardized Z_R score:

1. Start ArcView and use **File/Open Project** to open the `Ch3.apr` from `\AVStat\Chapter3\Scripts\`.
2. Click the **Add Theme** button and then navigate to the directory with `Ohcities.shp` and `Ohcounty.shp`. Then click the **OK** button.
3. Change the order of the two theme so that `Ohcities.shp` is at the top.
4. Click `Ohcities.shp` to make it active.
5. Choose **Point Patterns/Ordered Neighbor Statistics** from the menu bar in the **View** document.

6. In the *Set Properties* window, click the **Yes** button to continue.
7. In the *Nearest* window, click **OK** to continue. The default selection of **Nearest** is the first order neighbor relationship. Higher order neighborhoods can be accessed by clicking the appropriate entry from the drop-down list in this window.
8. Note that the calculated Observed Neighbor Distance is 8.51217. Click **OK** to continue.
9. The Expected Neighbor Distance shown in the next window is 11.6421. Click **OK** to continue.
10. The *R* value is shown in the **Nearest Neighbor Statistic:** window to be 0.731156. Click **OK** to continue.
11. The Z_R score is 6.58691, as shown in the next window. Click **OK** to continue.

As a result of this run, we found that the standardized Z_R score is very high, much higher than the 1.96 threshold we discussed earlier. We can conclude that the 164 cities are definitely clustered.

3.4   SPATIAL AUTOCORRELATION

In detecting spatial patterns of a point distribution, both Quadrat Analysis and Nearest Neighbor Analysis treat all points in the distribution as if they are all the same. These two methods analyze only the locations of points; they do not distinguish points by their attributes.

In this section, we will discuss a method for detecting spatial patterns of a point distribution by considering both the locations of the points and their attributes. This method uses a measure known as the *spatial autocorrelation coefficient* to measure and test how clustered/dispersed points are in space with respect to their attribute values. This measure is considered to be more powerful and more useful than the two methods discussed previously in certain ways. Different geographic locations rarely have identical characteristics, making it necessary to consider the characteristics of points in addition to their locations. Not only do locations matter; the conditions of these locations or activities happening there are also of great importance.

Spatial autocorrelation of a set of points is concerned with the degree to which points or things happening at these points are similar to other points or phenomena happening there. If significantly positive spatial autocorrelation exists in a point distribution, points with similar characteristics tend to be near each other. Alternatively, if spatial autocorrelation is weak or nonexistent, adjacent points in a distribution tend to have different characteristics. This concept corresponds to what was once called the *first law of geography* (Tobler, 1970): *everything is re-*

lated to everything else, but near things are more related than distant things (also cited and discussed in Gould, 1970, pp. 443–444; Cliff and Ord, 1981, p. 8; and Goodchild, 1986, p. 3).

With the spatial autocorrelation coefficient, we can measure

1. The proximity of locations and
2. The similarity of the characteristics of these locations.

For proximity of locations, we calculate the distance between points. For similarity of the characteristics of these locations, we calculate the difference in the attributes of spatially adjacent points.

There are two popular indices for measuring spatial autocorrelation in a point distribution: *Geary's Ratio* and *Moran's I*. Both indices measure spatial autocorrelation for interval or ratio attribute data. Following the notation used in Goodchild (1986, p. 13), we have

c_{ij} representing the similarity of point i's and point j's attributes,

w_{ij} representing the proximity of point i's and point j's locations, with $w_{ii} = 0$ for all points,

x_i represents the value of the attribute of interest for point i, and

n represents the number of points in the point distribution.

For measuring spatial autocorrelation, both Geary's Ratio and Moran's I combine the two measures for attribute similarity and location proximity into a single index of $\sum_{i=1}^{n} \sum_{j=1}^{n} c_{ij} w_{ij}$. It is used as the basis for formulating both indices. In both cases, the spatial autocorrelation coefficient (SAC) is proportional to the weighted similarity of attributes of points. Specifically, the equation for spatial autocorrelation coefficient takes the general form

$$SAC \approx \frac{\sum_{i=1}^{n} \sum_{j=1}^{n} c_{ij} w_{ij}}{\sum_{i=1}^{n} \sum_{j=1}^{n} w_{ij}}.$$

In the case of Geary's Ratio for spatial autocorrelation, the similarity of attribute values between two points is calculated as

$$c_{ij} = (x_i - x_j)^2.$$

The difference in attribute values for point i and point j is calculated as $(x_i - x_j)$. These differences for all pairs of i and j are then squared before being summed so that positive differences will not be offset by negative differences. Specifically, Geary's Ratio is calculated as follows:

$$C = \frac{\sum_{i=1}^{n} \sum_{j=1}^{n} c_{ij} w_{ij}}{2 \sum_{i=1}^{n} \sum_{j=1}^{n} w_{ij} \sigma^2} = \frac{\sum_{i=1}^{n} \sum_{j=1}^{n} w_{ij} (x_i - x_j)^2}{2 \sum_{i=1}^{n} \sum_{j=1}^{n} w_{ij} \sigma^2},$$

where σ^2 is the variance of the attribute x values with a mean of $\bar{x}$ or

$$\sigma^2 = \frac{\sum_{i=1}^{n}(x_i - \bar{x})^2}{(n-1)}.$$

In the case of Moran's I, the similarity of attribute values is defined as the difference between each value and the mean of all attribute values in question. Specifically, for Moran's I,

$$c_{ij} = (x_i - \bar{x})(x_j - \bar{x})$$

and the index can be calculated as

$$I = \frac{\sum_{i=1}^{n}\sum_{j=1}^{n} w_{ij}c_{ij}}{s^2 \sum_{i=1}^{n}\sum_{j=1}^{n} w_{ij}} = \frac{\sum_{i=1}^{n}\sum_{j=1}^{n} w_{ij}(x_i - \bar{x})(x_j - \bar{x})}{s^2 \sum_{i=1}^{n}\sum_{j=1}^{n} w_{ij}},$$

where s^2 is the sample variance or

$$s^2 = \frac{\sum_{i=1}^{n}(x_i - \bar{x})^2}{n}.$$

In Geary's Ratio and Moran's I, all terms can be calculated directly from the attribute values of the points. The only item not yet defined is w_{ij}, which is the proximity of locations between point i and point j. We often use the inverse of the distance between point i and point j. This assumes that attribute values of points follow the first law of geography. With the inverse of the distance, we give smaller weights to points that are far apart and larger weights to points that are closer together. For example, w_{ij} can be defined as $1/d_{ij}$, where d_{ij} is the distance between point i and point j.

The two indices are similar in format. The difference between them is whether the differences in attribute values (x_i and x_j) are calculated directly ($x_i - x_j$) or via their mean ($x_i - \bar{x})(x_j - \bar{x}$). As a result, the two indices yield different numeric ranges, as shown in Table 3.5. In Table 3.5, possible values for both

TABLE 3.5   Numeric Scales for Geary's Index and Moran's Index

Spatial Patterns	Geary's C	Moran's I
Clustered pattern in which adjacent points show similar characteristics	$0 < C < 1$	$I > E(I)$
Random pattern in which points do not show particular patterns of similarity	$C \sim 1$	$I \sim= E(I)$
Dispersed/uniform pattern in which adjacent points show different characteristics	$1 < C < 2$	$I < E(I)$

$E(I) = (-1)/(n-1)$, with n denoting the number of points in the distribution.

indices are listed with respect to three possible spatial patterns: clustered, random, and dispersed. Note that neither index's scale corresponds to our conventional impression of correlation coefficient of $(-1, 1)$ scale.

For Geary's Ratio, a value of 1 approximates the random pattern, whereas values greater than 1 suggest a dispersed or uniform pattern that has adjacent points displaying different characteristics. For Geary's Ratio, a value of less than 1 suggests a clustered pattern in which adjacent points show similar attribute values.

The numeric scale of Moran's I is anchored at the expected value of $E(I) = -1/n - 1$ for a random pattern. Index values that are less than $E(I)$ are typically associated with a uniform/dispersed pattern. At the other end of the scale, index values greater than $E(I)$ typically indicate clustered patterns where adjacent points tend to have similar characteristics.

When analyzing a point distribution, if we assume that the way the attribute values are assigned to the points is only one of the many possible arrangements using the same set of values, we adopt the assumption known as *randomization*, or *nonfree sampling*. Alternatively, we may assume that the attribute values in a set of points are only one of an infinite number of possibilities; each value is independent of others in the set of points. This assumption is sometimes called the *normality* or *free sampling* assumption. The difference between these two assumptions affects the way the variances of Geary's Ratio and Moran's I are estimated.

For both indices, we can calculate variances under free sampling and nonfree sampling assumptions. Free sampling allows replacements of observations in sampling possible outcomes, while nonfree sampling does not allow replacement.

Let's use R for the nonfree sampling assumption (randomization) and N for the free sampling assumption (normality). Following Goodchild (1986), we can estimate the expected values for a random pattern and the variances for Geary's C by

$$E_N(C) = 1$$

$$E_R(C) = 1$$

$$VAR_N(C) = \frac{[(2S_1 + S_2)(n - 1) - 4W^2]}{2(n + 1)W^2}$$

$$VAR_R(C) = \frac{(n - 1)S_1[n^2 - 3n + 3 - (n - 1)k]}{n(n - 2)(n - 3)W^2}$$

$$- \frac{(n - 1)S_2[n^2 + 3n - 6 - (n^2 - n + 2)k]}{4n(n - 2)(n - 3)W^2}$$

$$+ \frac{W^2[n^2 - 3 - (n - 1)^2 k]}{n(n - 2)(n - 3)W^2},$$

where

$$W = \sum_{i=1}^{n} \sum_{j=1}^{n} w_{ij}$$

$$S_1 = \frac{\sum_{i=1}^{n} \sum_{j=1}^{n} (w_{ij} + w_{ji})^2}{2}$$

$$S_2 = \sum_{i=1}^{n} (w_{i\cdot} + w_{\cdot i})^2$$

$$k = \frac{\sum_{i=1}^{n} (x_i - \bar{x})^4}{\left(\sum_{i=1}^{n} (x_i - \bar{x})^2 \right)^2}.$$

For Moran's I, the expected index value for a random pattern and the variances are

$$E_N(I) = E_R(I) = \frac{-1}{n-1}$$

$$VAR_N(I) = \frac{(n^2 S_1 - n S_2 + 3W^2)}{W^2(n^2 - 1)} - [E_N(I)]^2$$

$$VAR_R(I) = \frac{n[(n^2 - 3n + 3)S_1 - n S_2 + 3W^2]}{(n-1)(n-2)(n-3)W^2}$$

$$- \frac{k[(n^2 - n)S_1 - n S_2 + 3W^2]}{(n-1)(n-2)(n-3)W^2} - [E_R(I)]^2,$$

with W, S_1, S_2, and k similarly defined.

Once the expected values and their variances are calculated, the standardized Z scores can be calculated as

$$Z = \frac{I - E(I)}{VAR(I)}$$

or

$$Z = \frac{C - E(C)}{VAR(C)}.$$

Note that the same critical values of $-1.96 < Z < 1.96$ can be applied with a statistical significance level of 5%, or 0.05.

While the calculation of the spatial autocorrelation coefficient is straightforward, the definition of the similarity of locations can have some variations from those defined here. For example, w_{ij} can take a binary form of 1 or 0, depending on whether point i and point j are spatially adjacent. If two points are spatially adjacent, $w_{ij} = 1$; otherwise, $w_{ij} = 0$. If we use the concept of nodal region in the geography literature, each point in a distribution can be seen as the centroid of a region surrounding it. If the two adjacent regions share a common boundary, the two centroids of their corresponding regions can be defined as spatially adjacent.

Aside from various ways of defining w_{ij}, it is also possible to vary how the distance between points is used. For example, rather than defining $w_{ij} = 1/d_{ij}$, one can use $w_{ij} = 1/d_{ij}^b$ where b may take any appropriate value based on specific characteristics or empirical evidence associated with the geographic phenomena in question. This is because the distances measured by driving a car between two places can have quite different meaning from the distances measured by flying between two places or making phone calls between them. Many empirical studies indicated that $b = 2$ is widely applicable to many geographic phenomena.

Finally, it should be noted that spatial autocorrelation coefficients discussed here are also used for calculating similarity among polygon objects. We will discuss those uses in more detail in Chapter 5.

ArcView Notes The use of Geary's Ratio and Moran's I for a set of points in ArcView involves intensive computational time and a large amount of storage space on computer disks. This is because of the need to first calculate the distances between all pairs of points in the point theme. Depending on the hardware configuration and the amount of data, the computation will vary in the time it takes to finish calculating the distances. When working with a large data set, users should expect longer computational time than when working with a small data set.

The use of the spatial autocorrelation coefficient here is also included in the project file, Ch3.apr. Using the data set of the 17 northeast Ohio cities, the procedure for using Ch3.apr to calculate spatial autocorrelation is as follows:

1. Open Ch3.apr with ArcView and then add 5Ccities.shp and 5counties.shp.
2. Click the 5Ccities.shp theme to make it active.
3. From the **Point Patterns** menu item, choose **Create Distance Matrix** to start the process for computing the distances between points.
4. In the ID Selection window, select City_fips as the ID and then click **OK** to continue.
5. In the Output FileName window, navigate to your working directory and then click **OK** to save the file distmatrix.dbf.
6. Click **OK** in the next window to finish the processes of calculating distances.

The output file, `distmatrix.dbf`, contains the distances between all pairs of points in the point theme. If you wish, it can be added into the ArcView project by

1. highlighting the **Tables** icon in the project window,
2. clicking the **Add** button to invoke the Add Table window,
3. choosing the `distmatrix.dbf` file, where you stored it, and then clicking **OK** to add it to the current ArcView project, and
4. opening it as a regular ArcView Tables document.

With distance matrix computed, we are now ready to calculate index values for spatial autocorrelation by following these steps:

1. Choose the **Point Patterns** item from the menu bar.
2. In the drop-down submenu, choose **Moran-Geary** to proceed.
3. In the *Check for Input* window, answer the question "Have you created a spatial weight matrix?" by clicking **Yes** to proceed.
4. As previously specified, in the *Get Input* window, choose `City_fips` as the ID field. Click **OK** to continue.
5. In the next *Get Input* window, select `Pop1990` as the variable for this calculation. Click **OK** to continue.
6. For the next window, navigate to the directory where you have saved the `distmatrix.dbf` file. Choose it and then click **OK** to proceed.
7. In the *Weight* window, choose **Inverse to Distance** as the weight option for this calculation. Click **OK** to continue.
8. When the calculation is finished, the values for both indices are listed in the *Report* window (Figure 3.9) along with their expected values and variances under nonfree sampling (randomization) and free sampling (normality) assumptions. Note that the standardized *Z*-scores are also calculated for testing the statistical significance of the resulting index values.

In this example, we can see that the Moran's Index value is not statistically significant under either assumption. Geary's Index value, on the other hand, is statistically significant under the free sampling (normality) assumption.

3.5   APPLICATION EXAMPLES

The methods discussed in this chapter are used to detect spatial patterns of point distributions. Quadrat Analysis is concerned with how densities of points change over space, and it is a spatial sampling approach. Quadrats of consistent size and shape are overlaid on points in the study area. The frequency distribution of the

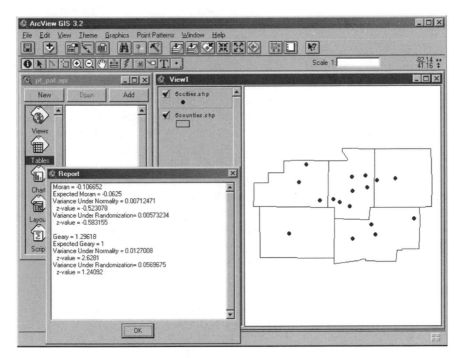

Figure 3.9   Reporting Moran's I and Geary's C Index values.

number of points in each quadrat is constructed and compared with the frequency distribution of a theoretical random pattern. Nearest Neighbor Analysis, on the other hand, exploits the spacing between neighbors. The distances between nearest neighboring points are measured. The average of the distances from all possible pairs of nearest neighbors is compared to that of a theoretical random pattern.

Both Quadrat Analysis and Nearest Neighbor Analysis are useful in detecting spatial patterns and comparing them with other known patterns. However, only the locations of the points are considered. These two methods do not take into account that different points in a distribution may be different in some ways or may represent different activities or events. Therefore, the use of these methods is limited.

Nevertheless, spatial autocorrelation coefficients consider the similarity of point locations as well as the similarity of attributes of the points. These coefficients calculate how attribute values change over space with respect to the locations of the points. Both Moran's I and Geary's C have been discussed in this chapter. They are equally useful but differ in their numeric scales.

In this section, we will look at how these methods can help us understand how point data distribute and, to a great extent, how we can use them to detect if the data distribute in any distinguishable pattern. We will use two sets of point-based data that represent the water transparency of monitored lakes.

The first data set concerns the transparency of lakes, as monitored by the Environmental Monitoring and Assessment Program (EMAP). EMAP is a program supported by the U.S. EPA that collects and analyzes data on environmental quality in the United States. The data used here are a subset of the data available from the program's website, which can be reached via the EPA's main web page. EMAP has developed an intensive monitoring plan that attempts to characterize fully the lake water quality of the northeastern United States. The data collected include transparency (as measured by the Secchi disk), water chemical variables, watershed characteristics, and information on fish, zooplanton, and diatom assemblages. The sampled lakes were selected using a stratified probabilistic approach, which randomly selected lakes based on a criterion that defined the statistical population of lakes in northeastern United States (Larsen et al., 1994).

Another data set is taken from the Great American Secchi Dip-In program. The North American Lake Management Society supports this program. Each year during the week of July 4th, thousands of volunteers across the United States and Canada dip their Secchi disks into lakes of their choice to measure the water transparency (Jenerette et al., 1998). These volunteers then report their findings to the program's office, along with their answers to other questions about the lake's environmental conditions and how the lake is being used. The selection of lakes being monitored has no prestructured framework. It is determined entirely by the volunteers. As a result, the lakes being monitored in this program represent the lakes that are being used, that volunteers care about, and consequently, the ones that need our attention.

One of the issues discussed recently is the sampling process used in the two lake monitoring programs. EMAP, through great efforts, selected lakes using what it considered to be a random pattern based on a stratified probabilistic approach. The Dip-In program, on the other hand, lets volunteers made the selections. The philosophical and theoretical approaches behind the two programs are entirely different, and it will be interesting to examine how the outcomes differ. To provide a visual impression of how the monitored lakes distribute, Figure 3.10 shows the locations of the EMAP and Dip-In lakes.

On the issue of how to better sample lakes to be monitored, we can use the methods discussed in this chapter to examine how the two data sets differ. We will measure to what degree the lakes being monitored by the two programs deviate from a random pattern to indicate indirectly how the sampling outcomes of the two programs differ.

Now that the ArcView project file, Ch3.apr, is available to us, it is just a matter of running the script for the two data sets. To examine the data in more detail, we can divide each data set by state boundaries to create subsets of data in both cases. This is done so that we can see how the spatial patterns change between scales. When the entire data set is used in the analysis, we test the spatial pattern at a multistate scale. When testing the subsets, we are examining the spatial patterns at a more detailed local scale.

Both data sets contain water transparency data (as an attribute in their ArcView shapefiles). Each of the data sets is divided into eight subsets for the follow-

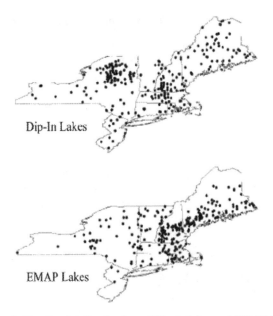

Figure 3.10   Spatial distribution of Dip-In lakes and EMAP lakes.

ing eight states: Connecticut (CT), Delaware (DE), Massachusetts (MA), Maine (ME), New Hampshire (NH), New York (NY), Rhode Island (RI), and Vermont (VT). Including the entire northeastern United States, each program has nine data sets in this analysis.

Table 3.6 lists the results of running Ch3.apr on each of the data sets and subsets. It gives the statistics and their Z scores for Quadrat Analysis and Nearest Neighbor Analysis. For Quadrat Analysis, 100 quadrats were used in each run.

In Table 3.6, the number of data points in each data set is shown in parentheses with the program's name. For example, EMAP has 350 lakes for the entire northeastern United States, and Dip-In has 303 lakes. Some subsets are excluded because one of the programs has fewer than five data points. They are being displayed with a gray screen. To identify significant results easily, those z-scores that are greater than 1.96 or less than -1.96 are highlighted in boldface italics because these z-scores indicate statistical significance at the $\alpha = 0.05$ (or 5%) level.

For the entire northeastern United States, neither program shows any spatial pattern when examined and tested by Quadrat Analysis. However, they are both considered nonrandom patterns by Nearest Neighbor Analysis. They deviate from the random pattern with a statistical significance at the $\alpha = 0.05$ level. When the data are partitioned into subsets for individual states, monitored lakes in both the EMAP and Dip-In programs show statistically significant dispersed patterns in Massachusetts, Maine, New Hampshire, and New York. For Vermont, EMAP's lakes show much more dispersion than the Dip-In lakes.

TABLE 3.6   Quadrat Analysis and Nearest Neighbor Analysis of EMAP Data and Dip-In Data

		Quadrat Analysis		Nearest Neighbor Analysis	
		D Statistics	*Z* score	NNA Statistics	*Z* score
NE	EMAP (350)	0.7505	0.0727	0.3227	*24.2433*
USA	Dip-In (303)	0.4308	0.0781	0.4128	*19.5540*
CT	EMAP (14)	*6.0000*	*0.3490*	*0.3103*	*4.9375*
	Dip-In (4)	*23.5008*	*0.6240*	*0.8198*	*0.6896*
DE	EMAP (13)	*6.3846*	*0.3610*	*0.7529*	*1.7041*
	Dip-In (4)	*23.0000*	*0.6240*	*0.4931*	*1.9397*
MA	EMAP (38)	1.5263	0.2206	0.4301	*6.7213*
	Dip-In (32)	2.0004	0.2200	0.6724	*3.5449*
ME	EMAP (74)	0.4553	0.1581	0.4416	*9.1909*
	Dip-In (99)	0.3456	0.1367	0.4612	*10.2563*
NH	EMAP (45)	1.1346	0.2027	0.5979	*5.1605*
	Dip-In (78)	0.4665	0.1540	0.4497	*9.2991*
NY	EMAP (142)	0.3486	0.1141	0.3403	*15.0407*
	Dip-In (44)	1.1818	0.2050	0.7324	*3.3958*
RI	EMAP (4)	*24.0008*	*0.6240*	*1.0870*	*0.3328*
	Dip-In (18)	*4.5556*	*0.3090*	*0.6070*	*3.1899*
VT	EMAP (18)	4.5000	0.3090	0.3119	*5.5853*
	Dip-In (21)	3.7156	0.2700	0.7998	*1.7556*

Notes:
• Numbers in parentheses are the number of points in each data set.
• Results from data sets with four or fewer points are screened.
• Z scores above 1.96 are highlighted in boldface italic.

Two observations can be made. First, Nearest Neighbor Analysis is a more powerful method than Quadrat Analysis because it detects what the Quadrat Analysis fails to detect. Second, volunteers, without special instructions, selected lakes that show dispersion similar to that of the lakes selected by EMAP's stratified sampling, but to a lesser extent.

When examining the data sets for spatial patterns, we often wonder what spatial pattern each data set will display. The spatial autocorrelation coefficients can be used to assist the detection. Table 3.7 shows the results of running Ch3.apr for calculating Geary's C index and Moran's I index. Similar to Table 3.6, states with fewer than five lakes in either the EMAP or Dip-In program are dropped from further analysis. Any z-score that is either greater than 1.96 or less than -1.96 is highlighted in boldface italic text, as it is statistically significant at the $\alpha = 0.05$ level.

With measures of water transparency data being tested by Geary's Ratio, EMAP shows some degree of a regionalized pattern but not much significance. Geary's Ratio is 0.9268 for the entire EMAP data set, suggesting that the spatial pattern shows smooth changes in water transparency between neighboring lakes.

TABLE 3.7   Spatial Autocorrelation In EMAP data and Dip-In Data

		Geary's C	Z score	Moran's I	Z score
NE	EMAP (350)	0.9268	0.0000	0.0279	*5.0743*
USA	Dip-In (303)	0.6814	−8.8528	0.1928	*16.3977*
CT	EMAP (14)	1.4593	1.9502	−0.3569	−1.4290
	Dip-In (4)	1.0253	0.3296	−0.3319	0.0043
DE	EMAP (13)	0.9383	−0.5924	0.0910	1.5382
	Dip-In (4)	0.0000	0.0000	0.0000	0.0000
MA	EMAP (38)	1.3206	0.0000	0.1774	*−2.8878*
	Dip-In (32)	0.8213	−1.0881	0.1120	1.1451
ME	EMAP (74)	0.9376	0.0000	−0.0082	0.1458
	Dip-In (99)	0.6876	−4.5359	0.0779	*3.0414*
NH	EMAP (45)	0.8728	−1.0937	0.1139	1.7596
	Dip-In (78)	0.7964	−3.6065	0.1474	*5.0601*
NY	EMAP (142)	0.8806	0.0000	0.0149	1.0342
	Dip-In (44)	0.9555	−0.4744	−0.0301	−0.1131
RI	EMAP (4)	1.0040	0.0404	−0.2987	0.0993
	Dip-In (18)	0.7976	−1.2640	−0.1070	−0.3794
VT	EMAP (18)	1.5770	1.9899	−0.4946	*−2.1601*
	Dip-In (21)	0.9505	−0.5391	−0.0184	0.3640

Notes:
- Numbers in parentheses are the number of points in each data set.
- Results from data sets with four or fewer points are screened.
- Z scores above 1.96 are highlighted in boldface italic.
- Z scores are calculated for the free sampling (or normality) assumption.

With the z-score approaching 0.0, the lakes monitored by EMAP do not show spatial autocorrelation with enough statistical significance. Also with Geary's Ratio, the lakes in the Dip-In program seem to show a much stronger degree of spatial autocorrelation. This means that the neighboring lakes tend to show similar values of water transparency. As suggested by the z-score, this data set may be more appropriate for analysis of a regional trend. When Moran's I is used, the lakes in both programs show a statistically significant departure from a random pattern. This is demonstrated by the high z-scores of 5.0786 (EMAP) and 16.3977 (Dip-In).

As for the data on individual states, the Dip-In program has Maine and New Hampshire showing strong regional trends, while Vermont shows contrasting transparency values between neighboring lakes. For the EMAP lakes, none of the states has a strong correlation of a point pattern to be detected by Geary's C index.

When using Moran's I index on data sets of individual states, we see that the EMAP program's lakes in Massachusetts and Vermont show strong dissimilarity between neighboring lakes in terms of water transparency. For Dip-In's data set, Maine and New Hampshire show a strong regional trend, as their transparency values tend to be similar between neighboring lakes.

With the example discussed in this section, we hope to show the different usages by the three methods. Before the availability of computer codes such as those in Ch3.apr, exploring various approaches to the analysis or detection of spatial patterns would have involved a tremendous effort. Now it is feasible and convenient.

REFERENCES

Boots, B. N., and Getis, A. (1988). *Point Pattern Analysis*. Newbury Park, CA: Sage Publications.

Cliff, A. D. and Ord, J. K. (1981). *Spatial Processes: Models and Applications*. London: Pion.

Gould, P. (1970). Is statistix inferens the geographic name for a wild goose? *Economic Geography*, 46: 439–448.

Goodchild, M. F. (1986). *Spatial Autocorrelation*. CATMOG 47. Norwich, CT: Geo Books.

Greig-Smith, P. (1952). The use of random and contiguous quadrats in the study of the structure of plant communities. *Annals of Botany* (London), New Series, 16: 312.

Griffith, D. A. and Amrhein, C. G. (1991). *Statistical Analysis for Geographers*. Englewood Cliffs, NJ: Prentice-Hall.

Jenerette, G. D., Lee, J., Waller, D., and Carlson, R. C. (1998). The effect of spatial dimension on regionalization of lake water quality data. In T. K. Poiker and N. Chrisman (eds.), *Proceedings of the 8th International Symposium on Spatial Data Handling*. I.G.U. G.I.S. Study Group. Burnaby, B.C., Canada: Inernational Geographical Union.

Larsen, D. P., Thornton, K. W., Urquhart, N. S., and Paulsen, S. G. (1994). The role of sample surveys for monitoring the condition of the nation's lakes. *Environmental Monitoring and Assessment*, 32: 101–134.

Taylor, P. J. (1977). *Quantitative Methods in Geography: An Introduction to Spatial Analysis*. Prospect Heights, IL: Waveland Press.

Tobler, W. R. (1970). A computer movie simulating urban growth in the Detroit region. *Economic Geography*, 46 (Supplement): 234–240.

CHAPTER 4

LINE DESCRIPTORS

In previous chapters, we have discussed how certain types of geographic features or events can be represented abstractly by points in a GIS environment such as ArcView. We have also discussed how locational information of these point features or events can be extracted from ArcView and utilized to describe their spatial characteristics. Furthermore, we have discussed the methods used to analyze them with other attribute data describing the features. In this chapter, let's shift our attention to the description and analysis of linear geographic features that can be represented most appropriately by line objects. We will describe two general linear features that can be represented in a GIS environment. Then, as in the previous chapter, we will discuss how geographic information can be extracted to study these linear features. Most of these analyses are descriptive.

4.1 THE NATURE OF LINEAR FEATURES

In a vector GIS database, linear features are best described as line objects. As discussed in Chapter 2, the representation of geographic features by geographic objects is scale dependent. For instance, on a small-scale map (1:1,000,000), a mountain range may be represented by a line showing its approximate location, geographic extent, and orientation. When a larger scale is adopted (1:24,000) or more detailed information is shown, a line is too crude to represent a mountain range with a significant spatial extent at that scale. At that scale, a polygon object is more appropriate. In other words, a geographic feature with significant spatial extent can be represented abstractly by a linear object at one scale but by a polygon at another scale. This process is sometimes known as *cartographic ab-*

straction. Another example is the Mississippi River and its tributaries. They have significant widths when they are shown on large-scale maps, but they are represented by lines on small-scale maps.

A line can be used to represent linear geographic features of various types. Most people and even GIS users use linear features for rivers and roads, but they actually can represent many more types of geographic features. Features of the same type within the same system are generally connected to each other to form a network. For instance, segments of roads connected together form a road network, and segments of streams belonging to the same river system or drainage basin form a river network or a drainage network of their own. Within these networks, individual line segments have their own properties. For example, each of them has a different length, different beginning and ending points, or different water/traffic flows. They are related to each other in a topological manner. These segments cannot be treated separately because of this topological relationship. Other examples include the network of power utility lines and the pipeline system of a gas utility.

Linear features, however, do not have to be connected to each other to form a network. Each of these features can be interpreted alone. For instance, fault lines of a geologically active area are probably noncontiguous. Other examples include spatially extensive features such as mountain ranges or touchdown paths of tornados. These spatially noncontiguous linear features can be analyzed as individual objects without any topological relationships.

Line objects in a GIS environment are not limited to representing linear geographical features (either networked or nonnetworked). They can also be used to represent phenomena or events that have beginning locations (points) and ending locations (points). For instance, it is quite common to use lines with arrows to show wind directions and magnitudes, which are indicated by the lengths of the lines. These are sometimes referred to as *trajectories*. Another example is tracking the movements of wild animals with Global Positioning System (GPS) receivers attached to them over a certain time period. In that case, the line objects represent where they started and where they stopped.

Figures 4.1–4.3 are examples of these three types of linear objects in GIS. Figure 4.1 shows a set of fault lines in Loudoun County, Virginia. Some of these fault lines are joined together for geological reasons. But topologically, there is no reason for these fault lines to be linked. In fact, Figure 4.1 shows many fault lines separately from other fault lines. In contrast to fault lines, which are geographic features, the linear objects shown in Figure 4.2 are events. The lines in Figure 4.2 show the trajectories of wind direction and speed (they can also be thought of as magnitude) at given locations. There are not geographic features we can observe, but there are geographic phenomena they can be represented. The line objects in Figure 4.2 represent the two common attributes of linear geographic features: direction and length. Different lines can have similar orientations (for instance, from north-northeast to south-southwest in Figure 4.1), but the directions of these lines can be opposite to each other. Therefore, an arrow is added to each line to show the direction of the wind. The length of the line can represent the spatial

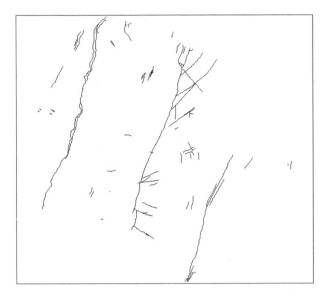

Figure 4.1   Selected fault lines in Loudoun County, Virginia.

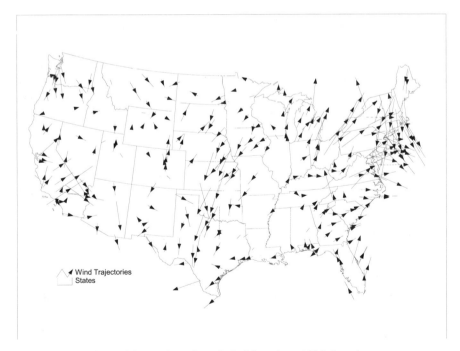

Figure 4.2   Trajectories of wind for selected U.S. locations.

Figure 4.3   Major road networks in the United States.

extent of the linear object, but in Figure 4.2, the length is proportional to the strength or magnitude of the wind. Figure 4.3 is a standard road network with line segments linking to each other. In addition to the characteristics of the other two types of linear features, a network of features shows how individual linear features are related topologically.

4.2 CHARACTERISTICS AND ATTRIBUTES OF LINEAR FEATURES

4.2.1 Geometric Characteristics of Linear Features

In most vector-based GIS, a linear object is defined either by a line segment or by a sequence of line segments that is sometimes referred to as a *chain*. If the object is relatively simple, such as the short, small fault lines shown in Figure 4.1, then a simple line segment will be adequate. But if the object is one side of the curb of a winding street, then several line segments may be required to depict its nonlinear nature. Similarly, a chain instead of a simple line will be rather effective and accurate in depicting an interstate highway.

If a line is used to represent a simple linear geographic feature, we would need two endpoints to define it. The locations of these two points may be defined in the form of longitude-latitude, x-y, or another coordinate system. Using the fault line as an example again, if we know that the fault line is short and a simple line segment is adequate to represent it, then all we need are the locations of the two points at both ends of the line. If a chain is required to depict a more complicated linear feature, in addition to the two endpoints of the chain, intermediate points depicting the sequence of line segments are needed to define the linear feature. The curb of a street is basically a collection of the edges of concrete blocks. Therefore, a line segment defined by two endpoints can represent the edge of a block, and a sequence of segments defined by two terminal points and a set of intermediate points can represent the curb. These chains can exist without linking to each other, like the fault lines in Figure 4.1. If chains are connected to each other, they form a *network*. In a network, linear objects are linked at the terminal points of the chains. A terminal point of a chain can be the terminal point of multiple chains, such as the center of a roundabout in a local street network or a major city such as St. Louis, where several interstate highways (I-70, I-55, I-64 and I-44) converge.

4.2.2   Spatial Attributes of Linear Features: Length

Linear features in GIS can carry attributes just like other types of feature. Here we focus only on spatial attributes that can be derived from linear features. To simplify the discussion, we can treat a simple line and a chain in the same manner in the sense that they can generally be defined by two terminal locations. In fact, most GIS made the intermediate points transparent to users. An obvious spatial attribute of any linear feature is its length. Given the locations of the two endpoints, the length of the linear feature can easily be calculated. After extracting the location of the two endpoints, we can apply the Pythagorean theorem to calculate the distance between the points and thus the length of the linear feature. The Pythagorean theorem states that for a right angle triangle (Figure 4.4), the sum of the squares of the two sides forming the right angle is equal to the square of the longest side. According to Figure 4.4, $a^2 + b^2 = c^2$. Therefore, if we know the x-y coordinates of the endpoints of c, then using the theorem, the length is

$$c = \sqrt{(x_1 - x_2)^2 + (y_1 - y_2)^2}.$$

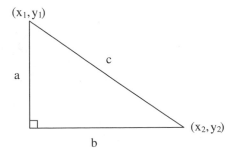

Figure 4.4   The Pythagorean theorem.

The above equation is appropriate to calculate the length of a simple line segment. If a set of line segments is linked together to form a chain, the length of the chain can be defined as the sum of these individual line segments if we are interested in the aggregated length of the chain. But if we are interested only in the spatial extent of the linear geographic feature, such as how far a river, which may be meandering, extends inland, then we can just calculate the straight-line length of the chain. This can be accomplished by taking the terminal locations of the chain and deriving the length between them as if a straight line is formed using the two terminal locations.

ArcView Notes

In the ArcView GIS environment, linear features are represented by the shape polyline. A polyline can be as simple as a simple line segment with only two endpoints. It can be more complicated, like a chain with multiple line segments concatenated together to depict more complicated geographic features. ArcView does support the simple line shape or object. But when geographic objects are stored in shapefiles, only the polyline shape is recognized.

Some spatial data sets used by ArcView may have length as an existing attribute. Otherwise, the information on length has to be extracted. Fortunately, in ArcView, we do not have to extract the locations of the endpoints and then apply the Pythagorean theorem in order to obtain the length.

The length of polyline shapes can be extracted either by using the **Field Calculator** with `.ReturnLength` request issued to the Polyline shape objects (please refer to Chapter 2) or by selecting the menu item **Add Length and Angle** in project file `Ch4.apr`. Either approach will put the length into the attribute table associated with the polyline theme in ArcView.

To use the **Field Calculator**, first choose **Start Editing** on the table under the **Table** menu. Under **Edit**, choose **Add Field** to add a field for the length. Please note that the `.ReturnLength` request provides the actual or true length of the polyline. Then use the **Field Calculator** to compute the length by inserting `[shape].ReturnLength` into the window.

If the polyline feature does not follow a straight line, the request will return the length following all the line segments when they sway from side to side. If the user is interested only in the length of the polyline feature marked by the two endpoints (straight line length), then the request `.AsLine.ReturnLength` should be used instead. This request first converts the polyline into a simple line segment defined by the two endpoints of the chain and calculates the length based upon the simplified line segment.

Another method used to extract and add the length information to the ArcView feature table is to use the new menu item created in the project file `Ch4.apr`. The interface in this project file is shown in Figure 4.5. Under the **Analysis** menu, a new item is **Add Length and Angle**. Choose this menu item to extract and add the length information of each polyline object to the feature table. The user will be asked to choose the straight-line length or the true length. Figure 4.6 shows part of the feature table after the two types of length information are extracted and added. The length attribute in the second column was originally provided in the table. Since the menu item **Add Length and Angle** was chosen twice, but the straight line length the first time and the true length the second time, both types of length were added to the table. It is clear that the first polyline feature, a long fault line in Figure 4.1, is a chain with segments going left and right. Thus, the *SLength* (for straight line length) is shorter than the *Tlength* (true length). The second polyline feature, however, is a rather short fault represented by a simple line segment. Therefore, *SLength* and *Tlength* are the same. After this step, the user can use the ArcView Statistics function to obtain simple descriptive statistics.

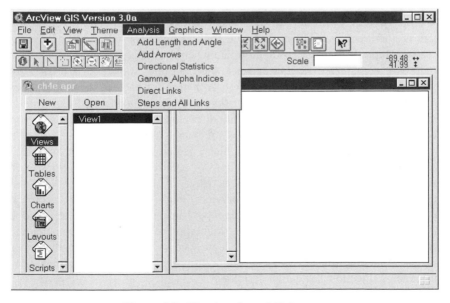

Figure 4.5   User interface of `Ch4.apr`.

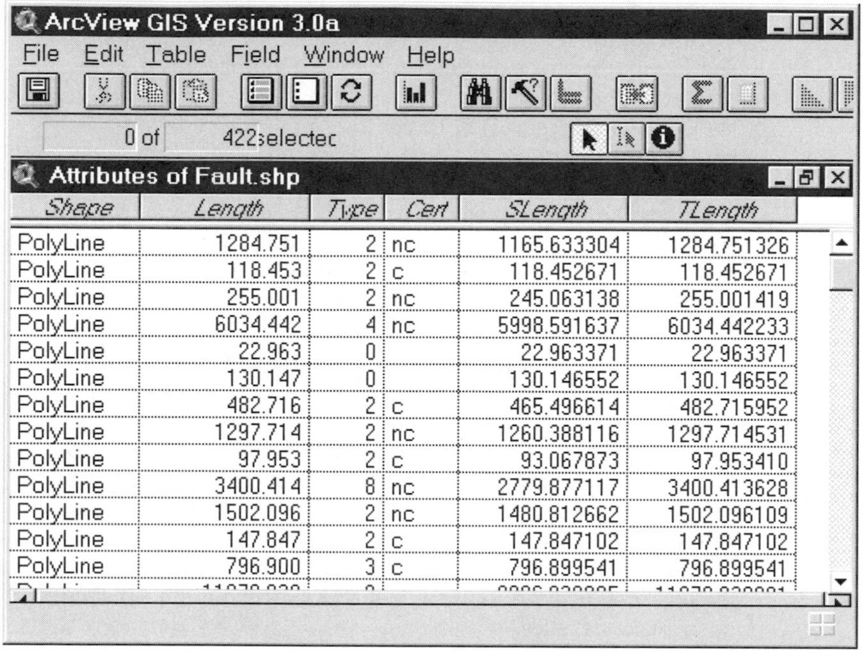

Shape	Length	Type	Cert	SLength	TLength
PolyLine	1284.751	2	nc	1165.633304	1284.751326
PolyLine	118.453	2	c	118.452671	118.452671
PolyLine	255.001	2	nc	245.063138	255.001419
PolyLine	6034.442	4	nc	5998.591637	6034.442233
PolyLine	22.963	0		22.963371	22.963371
PolyLine	130.147	0		130.146552	130.146552
PolyLine	482.716	2	c	465.496614	482.715952
PolyLine	1297.714	2	nc	1260.388116	1297.714531
PolyLine	97.953	2	c	93.067873	97.953410
PolyLine	3400.414	8	nc	2779.877117	3400.413628
PolyLine	1502.096	2	nc	1480.812662	1502.096109
PolyLine	147.847	2	c	147.847102	147.847102
PolyLine	796.900	3	c	796.899541	796.899541

Figure 4.6 Portion of feature table after length information is added.

4.2.3 Spatial Attributes of Linear Features: Orientation and Direction

Another obvious spatial attribute of a linear feature is its orientation. Orientation here is nondirectional. For instance, east-west orientation is the same as west-east orientation. Orientation is appropriate when the linear feature does not have a directional characteristic. For instance, the fault lines shown in Figure 4.1 are nondirectional. There is no specific *from-to* nature for each of those fault lines even though we can describe the fault line using *from location x to location y*. But if the fault line has orientation but not direction, then using *from* location *y to* location *x* to describe the fault line does not change its nature.

Other examples of nondirectional linear features include curbs, mountain ranges in small-scale displays, and sections of coastline. Usually, the orientation of a set of linear features can be indicated by verbal description, such as from north to south or from east to west, or vice versa. Another common approach is to use an angle, measured counterclockwise in reference to east, to describe the orientation precisely. Therefore, an orientation of 45 degrees means that the overall trend of the linear features is 45 degrees counterclockwise from the x-axis or east. Sometimes, however, the orientation may refer to the north instead. The referencing direction is situation dependent.

Arguing that direction is not appropriate for some linear geographic features does not imply that GIS data used to represent these nondirectional features do not record the directional information. In fact, most GIS capture the directions of linear features as the data are created even if the directions are not meaningful to the features. Depending on how the data are entered into GIS, quite often the beginning point and the ending point of a chain during the digitizing process define the direction of the chain. Therefore, the directions of fault lines, for example, are stored in the GIS data when the data are created even though direction is inappropriate to describe the fault lines.

With respect to orientation, a similar attribute of linear features is their direction. As implied in the discussion above, linear features have directional characteristics that are dependent on the beginning and ending locations. *From* location *x to* location *y* is not the same as *from* location *y to* location *x*. In fact, the directions of the two descriptions are exactly the reverse of each other, and the two descriptions can refer to two different linear features. For instance, a two-way street can be described as two linear features with exactly the same geographic location and extent but opposite directions. Linear objects representing events or spatial phenomena are often directional in nature. The wind trajectories described in Figure 4.2 are clearly of this type. Arrows are added to the lines to indicate the directions.

ArcView Notes In the ArcView GIS environment, it is not too easy to extract orientation or direction information from linear features, not quite similar to the extraction of length information. There is no request in ArcView to obtain the angle of a polyline. Based on the locations of the two terminal points, the orientation or angle of a linear feature can be obtained using trigonometric functions. In practice, this rather tedious process could be implemented using the **Calculate** function under the **Table** menu item. To simplify the procedure, we have developed a new menu item, **Add Length and Angle**, in the project file Ch4.apr to accomplish this task.

When that menu item is selected, users will be asked if they would like to add the directional information to the feature table. Then users will be asked if they would like to use orientation or direction. Users also have the choices of referencing to the east or the north in deriving the orientation or angle of direction. In order to demonstrate clearly the difference between orientation and direction, a simple make-up data set is used.

Figure 4.7 shows a simple three-polyline example. Arrows are added to the polylines to show their directions. Obviously, the three polylines have similar orientations. Two of them, however, have similar directions and the third has an almost opposite direction. Then **Add Length and Angle** was run four times to exhaust all four possibilities: orientation referenced to the east, orientation referenced to the north, direction referenced to the east,

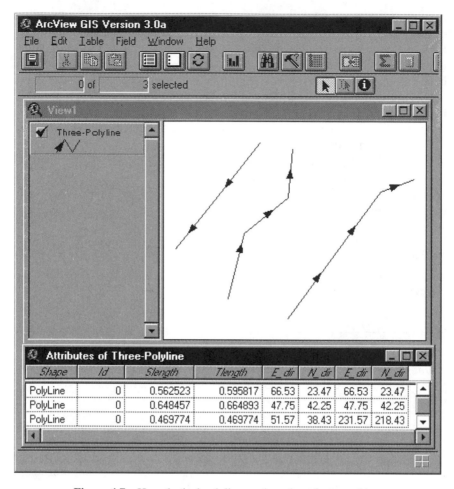

Figure 4.7   Hypothetical polylines and resultant feature tables.

and direction referenced to the north. The resultant feature table is also shown in Figure 4.7. These results confirm our visual inspections.

In the feature attribute table shown in Figure 4.7, the third and fourth columns present the length information. The fifth and sixth columns show orientation referencing to the east and north, and the seventh and eighth columns show angles of direction referencing to the east and north. The table clearly shows that the three polylines have similar orientations (between 48 and 67 degrees when referenced to the east). In regard to direction, the first two polylines have similar angles of direction (48 and 67 degrees when referenced to the east) but the third polyline has an opposite direction (232 degrees). Please note that for both orientation and direction, the angles are measured *counterclockwise* from the east and *clockwise* from the north.

4.2.4   Spatial Attributes of Linear Features: Topology

In a network database, linear features are linked together topologically. The attributes described in previous sections, such as length or spatial extent and orientation or direction, are also applicable to segments of the network. The length of the network, which can be defined as the aggregated lengths of individual segments or links, is an important feature in analyzing a network. Orientation or direction, depending on the specific natural of the network, is essential in understanding the geographic setting of the network. For instance, the direction of flow in tributaries of a river network should yield a consistent direction if the watershed is not very large.

The orientation of a major highway, to some extent, may partially reflect some characteristics of the landscape. Clearly, at a local scale, the direction of a local street network is important in planning the traffic pattern. All the concepts and analyses of linear features discussed so far are applicable in analyzing segments or links in a network. An additional aspect in a network, however, is how different segments are linked together and how these segments are related to each other. This is part of the general topic of the topological structure of a network.

The most essential topological aspect or attribute of a network is how different links or edges are connected to each other. This is sometimes known as the *connectivity* of a network. To capture quantitatively how different links are joined to other links, a traditional method is to use a matrix to store and represent the information. Assume that we have n links or edges in a network and each link has a unique identifier (ID). Conventionally, the labels of the columns are the ID numbers, and so are the labels of the rows in the *connectivity matrix*. The matrix is a square, that is, the number of rows equals the number of columns. A cell in the matrix captures the topological relationship between the two links denoted by the IDs in the corresponding row label and column label. If the two links are directly joined to each other, the cell will have a value of 1. Otherwise, the value will be 0.

Sometimes a link or an edge is not connected to itself, and therefore all diagonal elements in the matrix are 0s. Because each cell carries a value of either 0 or 1, this type of matrix is also called a *binary matrix*. And because the relationship between any pair of edges is symmetrical—that is, if Link A is connected to Link B, then Link B is also connected to Link A—the matrix is symmetrical. The triangle of matrix in the upper right side of the diagonal is a mirror of the lower left triangle.

Figure 4.8 shows a subset of major roads in the northeastern United States. All these roads in this example are in the state of Maine. Each road segment is la-

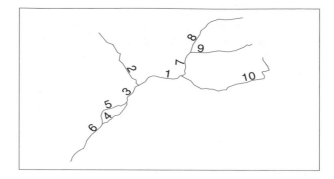

ID	1	2	3	4	5	6	7	8	9	10
1	0	1	1	0	0	0	1	0	0	1
2	1	0	1	0	0	0	0	0	0	0
3	1	1	0	1	1	0	0	0	0	0
4	0	0	1	0	1	1	0	0	0	0
5	0	0	1	1	0	1	0	0	0	0
6	0	0	0	1	1	0	0	0	0	0
7	1	0	0	0	0	0	0	1	1	1
8	0	0	0	0	0	0	1	0	1	0
9	0	0	0	0	0	0	1	1	0	0
10	1	0	0	0	0	0	1	0	0	0

Figure 4.8 A road network and its connectivity matrix.

beled with a unique ID. There are 10 road segments in this system. To capture the topological relationship among these segments, a connectivity matrix is created. Figure 4.8 also includes the connectivity matrix of the road network. The row labels and column labels refer to the IDs of the links or segments in the network. All diagonal elements, indicating whether each segment is connected to itself, are 0s.

In Figure 4.8, road segment 1 is in the middle of the network. It is connected to segment 2 to the northwest, segment 3 to the west, segment 7 to the northeast, and segment 10 to the east. Therefore, in the first row of the matrix, which indicates what road segments segment 1 is connected to, 1s are found in the cells for segments 2, 3, 7, and 10. The same principle is applied to other rows in the matrix. But if we focus on the first column, the information captured there is the same as that in the first row. Therefore, this is a symmetrical matrix.

Connectivity of links is the most fundamental attribute of a network. Any analysis of a network has to rely almost exclusively on the connectivity attribute. In fact, one may argue that connectivity defines a network. For instance, a set of linear features is shown on the screen of a GIS package. If we want to find the total length of that network, we may just add up the lengths of the individual segments appearing on the screen. But with very detailed visual inspection, we may find that two of those links that appeared to be joined together are in fact separated. Therefore, without topological information, any analysis performed on a network could be erroneous. In later sections, we will show how topological information serves as the basis of a suite of tools for network analysis.

We have briefly discussed attributes of linear features that are not necessarily connected and attributes of linear features that are connected topologically to form a network. To summarize, attributes that are applicable to linear features, but dependent on the nature and characteristics of these features, include length, orientation, direction, and connectivity. Please note that not all of these attributes are present or appropriate for all types of linear features. In the next section, we will discuss how different attributes of spatially noncontiguous linear features can be used to support various analytical tools specifically designed for these features. The following section will discuss tools we can use to analyze linear features in a network.

ArcView Notes In the project file for this chapter (`Ch4.apr`), a new menu item, **Connectivity Matrix**, is added for users to construct the connectivity matrix of a given network, as shown in Figure 4.8. The connectivity matrix will be a binary symmetric matrix in `dbf` format.

In some network or polyline databases, there may not be a unique identifier for each polyline in the ArcView feature table. The Connectivity Matrix function in the project file will ask the user if this situation exists. If there is no attribute item in the table that can be used as the unique identifier for each polyline, the new function will create a sequential unique identification number for each polyline to be used as the row and column labels in the connectivity matrix.

In other words, if there is no unique identifier in the feature table, but the user chooses one of the attributes as the identifier, the output may not be able to show the topological relationship among polylines accurately, as there is no way to distinguish one polyline from the others based upon the column or row labels.

In ArcView, any two objects will have a distance of 0 if any parts of the two objects touch each other. Using this characteristic in defining distance, the Avenue script for the connectivity matrix requests the distance between all pairs of polylines. If the distance between a given pair of polylines is 0, this means that they are connected; thus, a 1 will enter the corresponding cell in the matrix. Please note that we assume that the network structure follows the planar graph principle. That is, when two lines cross each other, a new vertex will be created; thus, the two lines are broken into four lines or segments. We do recognize, however, that two lines could cross each other without creating an intersection. For instance, a skyway over another highway or an overpass will be shown as two lines crossing each other, but in fact, topologically they do not intersect.

4.3 DIRECTIONAL STATISTICS

To analyze linear geographic features in greater depth, we have to rely on statistical techniques specifically designed for linear features. Unfortunately, not many statistical tools are available for analyzing linear features. Most of these techniques can be considered geostatistics developed and used mostly by geoscientists (Swan and Sandilands, 1995). Before we discuss these techniques, some preliminary and exploratory analyses can be conducted.

4.3.1 Exploring Statistics for Linear Features

In Section 4.1, we discussed the process used to extract some basic statistics or attributes of linear objects and to store the information in the feature attribute table as additional attributes. These statistics offer opportunities to conduct some preliminary and exploratory analyses. The length of a linear object—both the straight-line length and the truth length—can be analyzed using standard descriptive statistics such as mean, variance, and so on. In most situations, the analyses based upon these two length measures will probably yield slightly different results; however, the differences should not be dramatic. Table 4.1 shows two statistical summary tables of the fault line coverage shown in Figure 4.1. Part a

TABLE 4.1   Statistics Describing the Lengths of Fault Lines

(a) Statistics for the Slength Field	
Sum:	932,332.644656
Count:	422
Mean:	2,209.319063
Maximum:	39,005.08889
Minimum:	19.876572
Range:	38,985.212318
Variance:	16,315,314.639887
Standard deviation:	4,039.222034

(b) Statistics for the Tlength Field	
Sum:	648,470.093148
Count:	422
Mean:	2,247.559462
Maximum:	41,313.702596
Minimum:	19.876572
Range:	41,293.826024
Variance:	17,342,308.454853
Standard deviation:	4,164.409737

summarizes the attribute SLength, and Part b summarizes the attribute TLength. Conceptually, the two attributes should have the same value for a linear feature if the feature is represented by a simple line segment but not a chain. When a chain is needed, the straight-line length (SLength) will be shorter than the length of the entire chain (TLength). The two sets of summary statistics, including mean, sum, variance, and standard deviation, reflect these nature of the two attributes.

To go one step further, based on the difference between the straight-line length and the full length of the chain, we can analyze the topological complexity of each linear feature in the data set. One simple method is to derive a ratio of the two length measures—sinuosity (DeMers, 2000). When the length of the entire chain is divided by the length of the straight-line distance, the ratio is 1 if the linear feature is simple enough to be represented by a simple line segment. The higher this ratio is, the more complex the linear feature is. Table 4.2 shows selected fault lines with ratios larger than 1.2 in descending order. As a result, 10 fault lines have a ratio larger than 1.2. The first eight of them are shown in the top panel in Figure 4.9. All these faults are short but banned at one end, creating relatively large ratios for chain length to straight-line length. The other two fault lines on the list are shown in the middle panel of Figure 4.9. The ninth fault line looks like a curve, while the tenth is crooked. The ratio is very effective in identifying linear features that are relatively crooked.

Another attribute of a linear feature is direction. A simple exploratory method used to study linear features with a directional attribute is to add arrows to those features in a display to provide visual recognition of the pattern, if any. Figure 4.2, showing the trajectories of wind, provides a useful visual display of the phenomenon. Based upon the wind directions displayed in Figure 4.2, we can identify several circulation subsystems in the continental United States at that time. For instance, the New England region seemed to be influenced by one subsystem, while the Mid-Atlantic region throughout the South along the Atlantic coast seemed to be affected by another subsystem.

TABLE 4.2   Attribute Table for Selected Fault Lines with Length Ratios

Number	Length	Slength	Tlength	Length Ratio
1	596.841	386.532340	596.841079	1.54409
2	595.518	386.581012	595.517605	1.54047
3	698.879	483.387239	698.879055	1.44580
4	699.588	484.230785	699.558479	1.44580
5	651.754	469.902068	651.754268	1.38700
6	647.195	468.328944	647.194827	1.38192
7	603.488	441.679554	603.488095	1.36635
8	604.233	443.694169	604.232631	1.36182
9	3,400.414	2,779.877117	3,400.413628	1.22322
10	217.739	180.057803	217.738803	1.20927

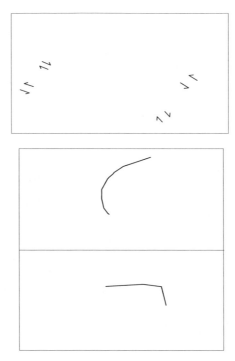

Figure 4.9   Selected fault lines with large true length to straight-line length ratio.

 ArcView Notes   In the `Ch4.apr` project file, a new menu item, **Add Arrows**, is added. With an active polyline theme in the **View** document, the **Add Arrows** menu item will add arrows to individual linear features to indicate their direction. Please note the following:

1. The arrows may be small, depending on the scale of the display in reference to the size of the features. Therefore, sometimes it is necessary to zoom in in order to see the arrows. An alternative method is to choose a larger point size for the arrows with the legend editor.

2. The direction of a linear feature is defined by the sequence of the points that are entered to define the feature if the data were created by digitizing. The direction of linear features, however, can be changed using requests in ArcView.

3. Arrows can be added to any linear features, but they may not be meaningful if the linear feature has an orientation attribute but not a directional attribute. As with fault lines in Figure 4.1, adding arrows to the feature is not meaningful.

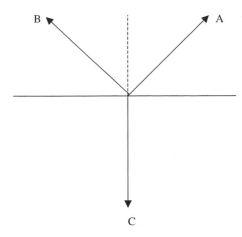

Figure 4.10   Inappropriateness of classic descriptive statistics for angle measures.

Since it is easy to analyze the length attribute of linear features, perhaps we can analyze the directional or orientation aspect of linear features with simple descriptive statistics. With the orientation and/or directional information of linear features extracted and added to the attribute data, we would naturally try to calculate descriptive statistics based on the angles (orientation or direction) of linear features. Unfortunately, descriptive classic statistics, such as mean or variance, are in general not appropriate to analyze angles of linear features. Figure 4.10 provides a simple example to illustrate this problem.

Figure 4.10 shows two vectors, A and B, with 45 degrees and 315 degrees, respectively, clockwise from the north. If we use the concept of mean in classical statistics to indicate the average direction of these two vectors, the mean of the two angles is 180 degrees, i.e., pointing south, as shown by vector **C**. But graphically, given the directions of vectors **A** and **B**, the average direction should be 0 degree (i.e., pointing north). Therefore, using classical statistical measures may be inappropriate. Because the concept of the arithmetic mean of the two angles cannot reflect the average direction, other measures such as variance cannot be defined meaningfully. To analyze angle information, we have to rely on directional statistics specifically designed to analyze vectors.

4.3.2   Directional Mean

The concept of *directional mean* is similar to the concept of average in classic statistics. The directional mean should be able to show the general direction of a set of vectors. Because directional mean is concerned with the direction but not the length of vectors, vectors can be simplified to vectors 1 unit in length (unit vectors). Figure 4.11a shows three vectors of unit length originated from O. Each vector shows a direction in reference to the origin. The directional mean is defined as the direction of the vector that is formed by "adding" all the vectors together.

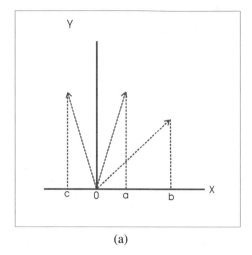

(a)

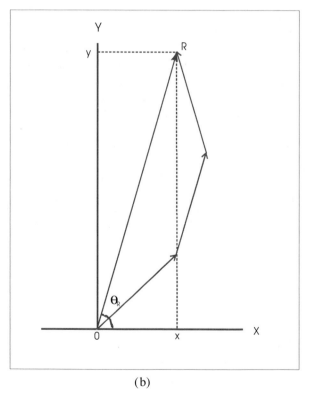

(b)

Figure 4.11 Concept of directional mean.

Adding any two vectors together means appending the beginning point of the second vector to the ending point of the first vector. Figure 4.11b shows how the three unit vectors in Figure 4.11a are added together. The result is the resultant vector, OR. The directional mean of the set of three vectors, in this case θ_R, is the direction of the resultant vector, OR. The direction of the resultant vector, θ_R, can be derived from the following trigonometric relation:

$$\tan \theta_R = \frac{oy}{ox}$$

where oy is the sum of the heights of the three vectors and ox is the horizontal extent of the vectors.

Because all three vectors are unit vectors, the height of a vector (in the y-axis) is basically the sine function of the angle of the vector, and the horizontal extent of a vector (x-axis, as shown in Figure 4.11a) is the cosine function of the angle of the vector. Therefore, if the three vectors are identified as a, b, and c, and their corresponding angles are θ_a, θ_b and θ_c, then

$$\tan \theta_R = \frac{\sin \theta_a + \sin \theta_b + \sin \theta_c}{\cos \theta_a + \cos \theta_b + \cos \theta_c}.$$

To generalize, assuming that there are n vectors v, and the angle of the vector v from the x-axis is θ_v, the resultant vector, OR, forms an angle, θ_R, counterclockwise from the x-axis. Because each vector is of unit length,

$$\tan \theta_R = \frac{\sum \sin \theta_v}{\sum \cos \theta_v},$$

which is the tangent of the resultant vector. In order to find the directional mean, the inverse of tan (arc tangent) has to be taken from the above equation.

The directional mean is the average direction of a set of vectors. The idea of vector addition, as shown in Figures 4.11a and 4.11b, utilizes the fact that the vector that results from adding vectors together shows the general direction of the set of vectors. If all the vectors have similar directions, after all vectors are appended to each other, the resultant vector will be pointing at somewhere among this group of vectors.

If two vectors have different directions, such as a vector of 45 degrees and a vector of 135 degrees, the resultant vector will be 90 degrees. As discussed before, just taking the average of angles of those vectors may not be appropriate. If all vectors under consideration are smaller or larger than 180 degrees, then arithmetically *averaging* the angles will yield the correct answer. However, if some angles are smaller than 180 degrees and others are larger, then the averaging method will be incorrect.

The result derived from the above equation for directional mean, however, has to be adjusted to accommodate specific situations in different quadrants, according to Table 4.3. The table shows the trigonometric results for angles in each of

TABLE 4.3   Adjusting the Directional Mean

sin $+ ve$ cos $- ve$	sin $+ ve$ cos $+ ve$
sin $- ve$ cos $- ve$	sin $- ve$ cos $+ ve$

the four quadrants. Because of these specific situations, the results from the calculation of directional mean should be adjusted accordingly:

1. If the numerator and the denominator are both positive in $\tan \theta_R$, no adjustment of the resultant angle is needed (the angle lies in the first quadrant).
2. If the numerator is positive and the denominator is negative (second quadrant), then the directional mean should be $180 - \theta_R$.
3. If both the numerator and the denominator are negative (third quadrant), then the directional mean should be $180 + \theta_R$.
4. If the numerator is negative and the denominator is positive (fourth quadrant), then the directional mean should be $360 - \theta_R$.

The directional mean can easily be computed if the angles of the linear feature are known. The first step is to compute the sine of each angle; the second step is to compute the cosine of each angle. Treating the sine and cosine results as ordinary numbers, we can calculate their sums to form the ratio in order to derive the tangent of the resultant vector. Taking the inverse of the tangent (arc tangent) on the ratio gives us the directional mean. Table 4.4 shows the major steps in deriving the directional mean of a selected numbers of fault lines described in Figure 4.1. First, the angles of the fault lines were extracted. Please note that directions of fault lines are not meaningful; their orientations are appropriate. Therefore, the direction mean is based upon orientation rather than direction.

The angles shown in Table 4.4 are orientations of fault lines from the east. First, the sines of the angles were derived, followed by their cosines. In many spreadsheet packages, trigonometric functions expect the angles expressed in radians instead of degrees. Therefore, the degrees of angles were first converted into radians before the sine and cosine functions were used. There are 422 fault lines in the data set. The sine and cosine values were summed. The ratio of the two sums is 1.7536, which is the tangent of the directional mean. Thus, taking the inverse tangent of 1.7536 gives us 60.31 degrees counterclockwise from the east.

4.3.3   Circular Variance

Analogous to classical descriptive statistics, *directional mean* reflects the *central tendency* of a set of directions. As in many cases, the central tendency, however, may not reflect the observations very efficiently. For instance, using the previous

TABLE 4.4   Directional Mean of Selected Fault Lines

Number	Length	E_DIR	sin_angle	cos_angle
1	1,284.751	68.56	0.930801	0.3655267
2	118.453	74.58	0.964003	0.2658926
3	255.001	94.09	0.997453	−0.071323
4	6,034.442	49.34	0.758589	0.651569
5	22.963	42.24	0.672238	0.7403355
6	130.147	43.6	0.68962	0.7241719
7	482.716	95.05	0.996118	−0.088025
8	1,297.714	89.74	0.99999	0.0045378
9	97.953	90.62	0.999941	−0.010821
10	3,400.414	57.28	0.841322	0.540534
⋮	⋮	⋮	⋮	⋮
414	2,976.265	58.1	0.848972	0.5284383
415	1,936.928	74.49	0.963584	0.2674066
416	10,528	74.19	0.96217	0.2724482
417	2,569.188	50.47	0.771291	0.6364822
418	167.585	70.12	0.949497	0.3400513
419	365.049	75.77	0.969317	0.245815
420	662.586	75.61	0.968627	0.2485208
421	274.119	77.3	0.975535	0.2198462
422	107.418	76.54	0.972533	0.2327665
		Sum =	301.914	172.76549

example, the directional mean of the two vectors, 45 degrees and 135 degrees, is 90 degrees. Apparently, the directional mean is not efficient in representing the two vectors pointing in very different directions. A more extreme case is one in which two vectors are of opposite direction. In this case, the directional mean will be pointing at the direction between them, but the statistic provides no information on the efficiency of the mean in representing all observations or vectors. A measure showing the variation or dispersion among the observations is necessary to supplement the directional mean statistic. In directional statistics, this measure is known as the *circular variance*. It shows the variability of the directions of the set of vectors.

If vectors with very similar directions are added (appended) together, the resultant vector will be relatively long and its length will be close to n if there are n unit vectors. On the other hand, if vectors are in opposite or very different directions, the resultant vector will be the straight line connecting the first point and the last point of the set of zigzag lines or even opposite lines. The resultant vector will be relatively short compared to n if for n vectors. Using the example in Figure 4.11, if all three vectors have very similar directions, the resultant vector, OR, should also be on top of the three vectors when they are appended to each other.

But in our example, the three vectors do vary in direction. Thus, OR deviates substantially from the graphical addition of the three vectors. As a result, OR is shorter than the actual length of the three vectors (which is 3). The length of the resultant vector can be used as a statistic to reflect the variability of the set of vectors. Using the same notations as above, the length of the resultant vector is

$$OR = \sqrt{\left(\sum \sin \theta_v\right)^2 + \left(\sum \cos \theta_v\right)^2}.$$

Circular variance, S_v, is $1 - OR/n$, where n is the number of vectors. S_v ranges from 0 to 1. When $S_v = 0$, OR equals n; therefore, all vectors have the same direction. When $S_v = 1$, OR is of length 0 when all vectors are of opposite directions, and the resultant vector is a point at the origin. Please note that the concept of circular variance is the same as the concept used when we compared the straight-line length of a chain with the length of the entire chain.

Using the fault lines as an example again,

$$\sum \sin 2_v = 301.914$$

$$\sum \cos 2_v = 172.765$$

$$\left(\sum \sin 2_v\right)^2 = 91,152.0634$$

$$\left(\sum \cos 2_v\right)^2 = 29,847.7452$$

$$OR = \sqrt{(91,152.0634 + 29,847.7452)} = 347.85$$

$$S_v = 1 - (347.85/422) = 0.1757.$$

Because circular variance ranges from 0 to 1, this magnitude of circular variance is rather small, indicating that most of the fault lines have similar directions.

In the next section, we will discuss techniques commonly used in analyzing a network.

ArcView Notes

In the Ch4.apr file, a new menu item, **Directional Statistics**, is already in place. In constrast to previous functions that either change or add information to the feature table (such as **Add Length and Angle**) or change the View (such as **Add Arrows**), this function alters neither the table nor the view. Instead, it provides information directly to the screen.

The procedure first asks users whether orientation or direction should be used. If orientation is chosen, then polylines are treated as lines in ArcView, and polyline directions will be either from lower left to upper right or from lower right to upper left. If direction is important, the directions inherited in the polylines will be used. The procedure will then derive the directional mean. Then the circular variance is calculated.

Using the wind trajectories in Figure 4.2 as an example, the **Directional Statistics** function is used. Because the trajectories are directional, when the **Directional Statistics** function was executed, direction was selected as important. The directional mean (counterclockwise from the east) for all the vectors is 356.54 degrees, which is an east-southeast direction. Figure 4.2 shows that there were at three major wind subsystems in the continental United States at that time. We can divide the entire region into three subregions: east, central, and west. The overall wind direction in the east is apparently pointing toward the east or slightly toward the northeast. Wind in the central area was blowing strongly toward the south or southwest. The west is a bit chaotic, but in general the directions seem to be pointing toward the southeast. Before the **Directional Statistics** function was executed, vectors in each of these three regions were selected. The results reported in Figure 4.12 confirm our speculations.

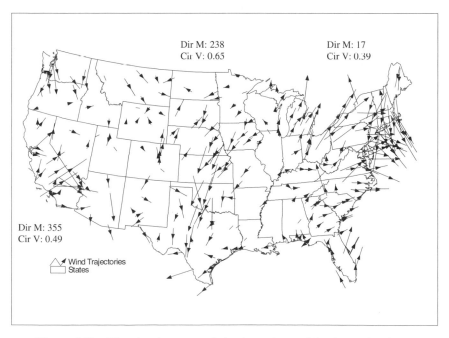

Figure 4.12   Directional means and circular variance of the three subregions.

The results of circular variance (Figure 4.12) provide additional information on wind direction in the three subregions. According to the circular variance results, the eastern section has a rather uniform direction, while the central region has the most varied wind direction. The variation of wind directions in the west is between those of the east and the south. Therefore, the spatial selection tools in GIS can be combined with the **Directional Statistics** function to analyze regional variations.

4.4 NETWORK ANALYSIS

To conduct a full range of network analyses, a special computation environment is required. The current version of ArcView cannot support this environment without extensive programming development. In addition, the Environmental System Research Institute (ESRI) has provided Network Analyst, an extension for Arc-View, to support certain types of network analysis. Therefore, we will discuss topics in network analysis that will not significantly overlap with the literature relying on the Network Analyst extension.

We have already discussed the basic attribute of a network—*connectivity*. Connectivity information can be captured in a binary symmetric matrix, indicating the pairs of links or segments that are joined together. This information will be used in almost all analyses involving a network (Taaffe, Gauthier, and O'Kelly, 1996). In general, we can classify different types of network analytical tools into two categories:

- The type that accesses the overall characteristics of the entire network

- The type that describes how one network segment is related to other segments or the entire network system

The first type will be discussed in the next section on connectivity; the second type, accessibility, will be covered in Section 4.4.2.

Before we start, we have to define several terms that are used in network analysis. In network analysis, a segment of linear feature is called a *link* or an *edge*. An edge is defined by the two vertices or nodes at both ends of the edge in the network. The number of edges and vertices are often used to derive statistics indicating the characteristics of the network.

4.4.1 Connectivity

So far, we have discussed only how the connectivity information is captured and represented in the connectivity matrix. We have not analyzed the level of connectivity in a network. For a fixed set of vertices, different networks can be created if the vertices are connected differently. When the number of vertices is fixed, networks with more edges are better connected. There is a minimum number of edges that is required to connect all the vertices to form a network.

If v denotes the number of vertices and e denotes the number of edges in the network, then the minimum number of edges required to link all these vertices to form a network is

$$e_{min} = v - 1.$$

In a *minimally connected network*, if any one edge is removed from the system, the network will be broken up into two unconnected subnetworks. In the simple network shown in Figure 4.13, there are 10 vertices and 10 edges. Since the number of vertices is 10, a minimally connected network should have 9 (10−1) edges. In this case, however, there are 10 edges. Therefore, the network in Figure 4.13 is not a minimally connected network. If either edge 4 or 5 is removed, then this is a minimally connected network.

Similarly, given a fixed number of vertices, there is a maximum number of edges one can construct to link all the vertices together. The edges in the maximally connected network do not cross or intersect each other (the *planar graph topology*). The maximum number of edges possible in the network is

$$e_{max} = 3(v - 2).$$

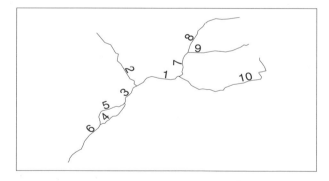

ID	1	2	3	4	5	6	7	8	9	10
1	0	1	1	0	0	0	1	0	0	1
2	1	0	1	0	0	0	0	0	0	0
3	1	1	0	1	1	0	0	0	0	0
4	0	0	1	0	1	1	0	0	0	0
5	0	0	1	1	0	1	0	0	0	0
6	0	0	0	1	1	0	0	0	0	0
7	1	0	0	0	0	0	0	1	1	1
8	0	0	0	0	0	0	1	0	1	0
9	0	0	0	0	0	0	1	1	0	0
10	1	0	0	0	0	0	1	0	0	0

Figure 4.13   A simple network in Maine.

Therefore, the network in Figure 4.13 can have up to 24 edges $(3 * (10 - 2))$ for the given number of vertices in the network. But actually, the network has only 10 edges. Thus, the Gamma Index (γ), which is defined as the ratio of the actual number of edges to the maximum possible number of edges in the network, is

$$\gamma = \frac{e}{e_{max}}.$$

The Gamma Index for the network in Figure 4.13 is 10/24, which is 0.4167. The Gamma Index is most useful in comparing different networks to differentiate their levels of connectivity. To illustrate this aspect of the index, another network selected from the highway network in northern New Jersey is shown in Figure 4.14. It is easy to tell by visual inspection that the New Jersey network is much better connected than the highway network in Maine. To verify this conclusion, we calculate the Gamma Index. In the New Jersey network, there are 17 vertices and 21 edges. Therefore, the Gamma Index is 0.4667, higher than that in the Maine network.

Another characteristic of connectivity is reflected by the number of circuits that a network can support. A *circuit* is defined as a closed loop along the network. In a circuit, the beginning node of the loop is also the ending node of the loop. The existence of a circuit in a network implies that if it is for a road network, travelers can use alternative routes to commute between any two locations in the network. A minimally connected network, as discussed above, barely links all vertices together. There is no circuit in such a network. But if an additional edge is added

Figure 4.14   A partial highway network in northern New Jersey.

to a minimally connected network, a circuit emerges. Therefore, the number of circuits can be obtained by subtracting the number of edges required for a minimally connected network from the actual number of edges. That is, $e - (v - 1)$ or $e - v + 1$. In the network shown in Figure 4.13, with $e = 10$ and $v = 10$, the number of circuits is 1. This is formed by edges 4 and 5. For a given number of vertices, the maximum number of circuits is $2v - 5$. Therefore, with these two measures of a circuit, we can derive a ratio of the number of actual circuits to the number of maximum circuits. This ratio, which sometimes is known as the *Alpha Index*, is defined as

$$\alpha = \frac{e - v + 1}{2v - 5}.$$

Using the Alpha Index, we can compare the two networks in Figures 4.13 and 4.14. In the network in Maine, if not for the circuit formed by edges 4 and 5, this network would be a minimally connected network. Thus, the Alpha Index of this network is only 0.067. By contrast, the network in New Jersey is better connected, with several circuits identifiable through visual inspection. The Alpha Index turns out to be 0.172, more than twice as high as the index for the Maine network.

ArcView Notes In the project file `Ch4.apr`, a new function is added to the standard ArcView user interface to calculate the two network indices. The procedure first identifies the number of edges and the number of vertices in the network. ArcView, which does not utilize the arc-node topology in the data, as in ARC/INFO, does not offer simple methods to identify nodes or vertices in the network. The procedure basically decomposes all polylines into points in order to identify the endpoints of each polyline. By comparing the endpoints of different polylines, the procedure identifies the number of vertices. On the basis of the number of vertices and the number of edges, the two indices are derived.

4.4.2 Accessibility

The two indices discussed above assess the characteristics of the entire network. Different individual elements, either vertices or edges, have different characteristics or relationships throughout the network. Therefore, it is necessary to analyze the characteristics of each of these elements. In general, the procedure is to evaluate the accessibility of individual element in regard to the entire network.

Traditionally, network analysis literature uses vertices or nodes as the basic elements of analysis (Taaffe, Gauthier, and O'Kelly, 1996). Conceptually, using links or edges as the basic elements is also appropriate because the accessibility results are now applied to the edges but not the vertices. In the rest of this chapter, our discussion will focus on the accessibility of individual edges.

A simple analysis is to identify how many edges a given edge is directly connected to. If the edge is well connected, then many edges will be linked to it directly. For instance, in the network in Maine, link 1 is well connected because it has four direct links to other edges. Links 3 and 7 have the same level of accessibility. On the other hand, Links 2, 6, 8, 9 and 10 share the same level of accessibility based upon the number of direct links. To a large degree, this information is already captured in the connectivity matrix such as the one shown in Figure 4.13. The binary connectivity matrix shows a 1 if the corresponding edges are directly linked and 0 otherwise. Therefore, in order to find out how many direct links are established for an edge, we just need to add all the 1s across the columns for the corresponding rows. Table 4.5 shows the connectivity matrix with the sum of direct links for each edge. The results confirm our earlier analysis based upon the map in Figures 4.8 and 4.13.

However, the number of direct links to an edge may not accurately reflect the accessibility of an edge. An edge with a small number of direct links may still be quite accessible, depending on its relative (topological) *location* in the network. For instance, an edge may have a small number of direct links but may be centrally located in the network, so that it is still well connected to other edges. Edge 2 in Figure 4.13 is such a case. It has only two direct links (to Links 1 and 3), but it is centrally located. Thus, it is easy to travel from this edge to all other edges in the network as compared to Link 6, which is located at the end of the network. To capture the relative location of a given edge in the network, we need to find out how many links are required to reach the farthest part of the network from that edge.

In a simple network like the one in Maine, it is quite easy visually to derive the number of links or steps required to reach the farthest edges. Using the edges identified, for Link 1, the most accessible one according to the number of direct links, three links or steps are required to reach the farthest edge (Link 6). Link 2, one of those with the lowest number of direct links, also requires three steps to reach the farthest edges (Links 6, 8, and 9). Therefore, even if the two edges have

TABLE 4.5 Connectivity of the Network in Maine and the Number of Direct LInks

ID	1	2	3	4	5	6	7	8	9	10	Sum Links
1	0	1	1	0	0	0	1	0	0	1	4
2	1	0	1	0	0	0	0	0	0	0	2
3	1	1	0	1	1	0	0	0	0	0	4
4	0	0	1	0	1	1	0	0	0	0	3
5	0	0	1	1	0	1	0	0	0	0	3
6	0	0	0	1	1	0	0	0	0	0	2
7	1	0	0	0	0	0	0	1	1	1	4
8	0	0	0	0	0	0	1	0	1	0	2
9	0	0	0	0	0	0	1	1	0	0	2
10	1	0	0	0	0	0	1	0	0	0	2

TABLE 4.6   Number of Direct Links, Steps Required to Reach the Farthest Part of the Network, and Total Number of Direct and Indirect Links for Each Edge

ID	Link	Steps	All_Links
1	4	3	15
2	2	3	19
3	4	3	16
4	3	4	21
5	3	4	21
6	2	5	28
7	4	4	18
8	2	5	25
9	2	5	25
10	2	4	20

very different numbers of direct links, their locations in the network are equally desirable. Table 4.6 shows the number of direct links and the number of steps or links required to reach the farthest part of the network. The two measures do not necessary yield the same conclusion because they evaluate different aspects of the edges with respect to other edges. Please note that 1 plus the highest number of links or steps required to reach the farthest edge of the entire network is also known as the *diameter* of the network. In Figure 4.13, the highest number of steps required to reach the farthest edge of the network is five, which is defined by Links 6, 8, and 9. Therefore, the diameter of the network is 6, i.e., six edges are required to link the farthest parts of the network.

An analysis based solely on the number of direct links is not the most reliable. An edge may not have many direct links, but because of its location, it may be reached from other edges indirectly and thus may be reasonably accessible. Therefore, besides direct linkages, our analysis should take indirect linkages into consideration. But obviously, a direct link is better than an indirect link; therefore, the two types of links should be treated differently. Indirect links also have different degrees of connectivity. For instance in the network in Maine, Links 1 and 8 are indirectly linked. Because Link 7 is between them, this indirect link is inferior to a direct link. Links 2 and 8 are also indirectly linked. But between them are Links 1 and 7. Therefore, two steps are required to join Links 2 and 8, one more step than the indirect link between Links 1 and 8. Thus, using the number of steps between links can indicate the quality of the link. If more steps are required, the indirect link is less desirable.

On the basis of this idea, we can derive the number of direct and indirect links that will require all edges to be joined together for each given edge, but the indirect links will be weighted by the number of steps or the degree of indirectness. Apparently, the larger the number of these total links, the less accessible is the edge. A small number of total links implies that the edge is well connected directly and indirectly to other edges. Table 4.6 also includes a column of all links for each edge. Link 6 requires the largest number of direct and indirect links in order to connect the entire network; thus, it may be regarded as the most inaccessible edge. This result conforms with the results of the number of direct links and the maximum number of steps required to reach the farthest part of the network. However, the total number of links is better in differentiating Link 6 from Links 8 and 9. The latter two edges have the same number of direct links and steps as Link 6, but in terms of the total number of links, Link 6 is a little less efficient than the other two edges. Similarly, the other two measures cannot distinguish the accessibility between Links 1 and 3, but the total number of links shows that Link 1 is slightly more accessible than Link 3.

ArcView Notes Project file `Ch4.apr` includes two new menu items: **Direct Links** and **Steps and All Links**. The first new function adds the number of direct links for each polyline to the feature table. The second new function adds the other two measures of accessibility to the feature table in ArcView. These two measures are the number of links or steps required to reach the farthest part of the network and the total number of direct and indirect links. In deriving the number of direct links, the procedure is identical to the procedure creating the connectivity matrix. The additional steps consist of adding the number of direct links for each edge and inserting them in the feature table in ArcView instead of the connectivity matrix.

To derive the number of links or steps required to reach the entire network, the procedure identifies direct links of a link or a selected set of links in a recursive manner until all edges are exhausted. To derive the total number of direct and indirect links, indirect links are given weights according to the number of steps. If the indirect link involves three edges, it has a weight of 3. All these weights are added together to show the total number of direct and indirect links.

Because these measures of accessibility are added to the feature table, they can be shown by maps. Figure 4.15 shows three maps for the three measures for the road network in Maine. These maps are a very effective graphic representation of the accessibility analysis. They are very informative in showing the relative levels of accessibility for different edges in the network.

Please note that the accessibility algorithms implemented here are for demonstration purpose only. Using especially the last function on large networks (more than 100 segments) may require extensive processing time. Traditional network analysis relies heavy on the connectivity matrix. Using various matrix manipulations and algebra, many results described in this chapter can be derived. Most GIS including ArcView, however, do not

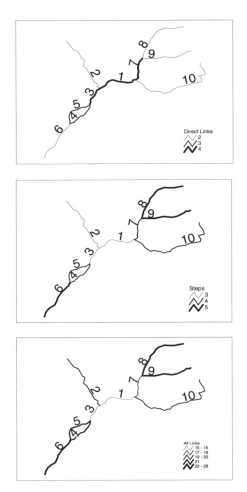

Figure 4.15   Maps showing the three measures of accessibility.

support advanced mathematical formulation and computation; therefore, the matrix approach for network analysis was not implemented here. The approach developed here is a noncomputational GIS approach exploiting the characteristics and behavior of GIS objects.

4.5   APPLICATION EXAMPLES

In this chapter, we have discussed several descriptive tools for analyzing linear features. These linear features may be completely or partially connected, such as fault lines or streams. The geometric characteristics analyzed in this chapter include their straight-line lengths, their network lengths, their overall directions or orientations, and consistency in their directions or orientations. These linear features can be fully connected, such as the road network of a given area or all tributaries of a river. Besides analyzing their geometric characteristics as if they are not fully connected, we can analyze how well they are connected to each other. We have discussed indices for the overall connectivity of the network and indices showing the relative accessibility of different network segments. In the balance of this chapter, we present two examples to demonstrate how the tools introduced in this chapter can be used.

4.5.1 Linear Feature Example

Using the drainage data for the northwestern corner of Loudoun County, Virginia, as an example, we will show how some of the tools we have discussed can be used to analyze a simple phenomenon. The terrain of northwestern Loudoun County is relatively rugged. On visual examination of the stream network, the drainage system appears to be divided by the mountain ridge running from north to south. Figure 4.16 shows the streams in that part of the county. It is clear that a long strip from north to south is clear of all streams and divides the system into two subsystems. If this assumption is correct and we have no additional geologic information, we may conclude that the streams on the two sides of the ridge probably have different directions.

Before we analyze the directions of the potential subsystems, we should analyze other geometric characteristics of the streams. Altogether there are 2,206 line segments in the network. Originally, there was no length information for the

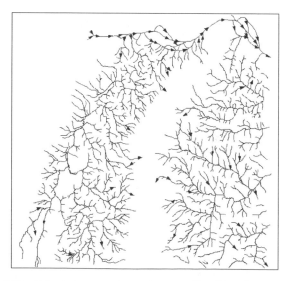

Figure 4.16 Selected streams in northwestern Loudoun County, Virginia.

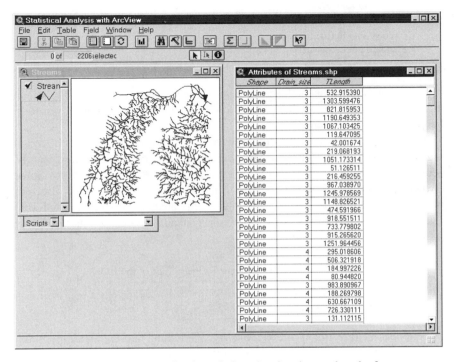

Figure 4.17 ArcView application window showing the true length of streams.

streams in the feature attribute table. Using the new function **Add Length and Angle** in the project file (ch4.apr) for this chapter, we can include the length of each stream in the feature table. We can also calculate the direction and orientation of each stream, but for this example, these measures will not be very useful. Figure 4.17 shows the application of ArcView with the feature table of the stream database. The table includes the true length of each stream segment.

To analyze the lengths of streams on the two sides of the ridge, we used the following procedure:

1. Using the selection tool in ArcView, select all streams on the west side of the ridge.
2. Open the feature attribute table of the streams, as shown in Figure 4.17, and highlight or select the field *TLength*.
3. Using the **Statistics** function in the table, derive the set of statistics.
4. Repeat the above steps for the east side of the ridge.

Below are the comparison of two statistics we found for the two sides of the ridge:

	West Side of the Ridge	East Side of the Ridge
Average length (m)	376.58	411.30
Standard deviation (m)	355.85	365.23

It seems that the streams on the east side of the ridge, on average, are longer than those on the west side, but the streams on the east side are also more varied in length.

In terms of direction of the streams, we could add an arrow to each stream segment using the new function **Add Arrow** in the project file. Unfortunately, because the streams are very dense, not all arrows appeared in the view. Some arrows did appear in Figure 4.16. Still, a more formal analysis of the direction is warranted. Using the procedure similar to the analysis of stream length, but selecting the **Directional Statistics** function instead, we obtained the following results:

	West of the Ridge	East of the Ridge
Directional mean	312.805	325.995
Circular variance	0.4536	0.4456

Note that the directional means are derived assuming that the direction option is selected. That is, the direction of flow of the stream is considered. Because both directional means are between 270 and 360 degrees, they point to the southeastern direction. Overall, streams on the east side of the ridge orient toward the east more than streams on the west side. Streams on both sides seem to vary in direction by a similar magnitude. But overall, we cannot say that streams on both sides of the ridge run in different directions.

One has to realize that main streams and tributaries run in different directions and sometimes are even orthogonal to each other. The above analysis did not distinguish streams of different sizes; therefore, the results may not be accurate. Using ArcView selection and query tools, we select larger streams (using the **Drain-size** attribute in the feature table to select **Drain-size 1** or **2** with 882 streams selected) from the entire region. The selected streams are shown in Figure 4.18. We performed the same analysis as above and achieved the following results:

	West of the Ridge	East of the Ridge
Directional mean	314.928	336.01
Circular variance	0.4147	0.38981

When the analysis is limited to larger streams, the directions on the two sides of the ridge diverge a bit more. A significant change occurs in the consistency of stream directions. The circular variances for both sides are smaller than before, and the decline in circular variance is very clear on the east side. These results indicate that streams on the east side flow more to the east, while streams on the west side flow more to the south-southeast.

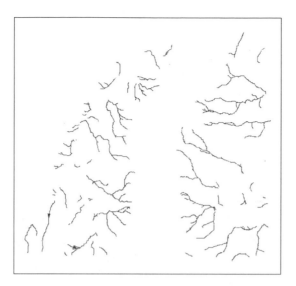

Figure 4.18   Selected larger streams, Loudoun County, Virginia.

4.5.2   Network Analysis Example

The other analytical tools introduced in this chapter are used for analyzing networks. Two data sets are selected as examples to demonstrate how these tools can be used. The Washington, DC, metropolitan area has one of the worst traffic conditions of any major U.S. cities. As the capital of the country, this city should have a very effective road network. Washington, DC, is surrounded by several local jurisdictions. Several Maryland counties are on the east and north, and several Virginia counties are on the west and south. The worst traffic areas are found in Montgomery County, Maryland, and several counties in northern Virginia, including Fairfax (including the independent city of Fairfax in the middle), Arlington, and the City of Alexandria. Figure 4.19 shows the boundaries of these jurisdictions, their road networks, and Washington, DC. The major road network data are divided into two groups: one for Montgomery County and the other for northern Virginia. The data set includes only major roads such as interstate highways and local major highways. Local streets are excluded for illustrative purposes. We will use the network analytical tools discussed in this chapter to analyze and compare these two road networks.

The following are the steps we took:

1. Move the Road Theme for northern Virginia to the top of the Table of Contents in the View window to ensure that the network analysis procedures will use that theme.
2. Under the **Analysis** menu, choose the menu item **Gamma_Alpha Indices**.

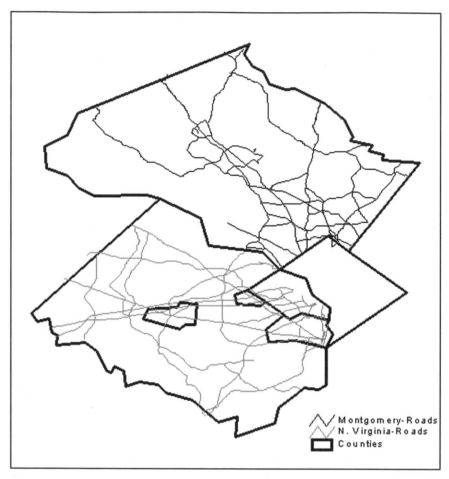

Figure 4.19   Selected road networks for northern Virginia and Montgomery County, Maryland.

3. If no feature is selected, ArcView will notify the user and ask the user if it is all right to use all features in the theme.
4. After clicking **OK**, a series of measures shows up in the window.
5. Move the Road Theme of Montgomery County to the top of the Table of Contents in the View window, and repeat steps 2 to 4 for Montgomery County.

The results from the two areas are reported in Table 4.7. It seems that northern Virginia has a denser major road network than Montgomery County, based upon the vertex numbers and numbers of edges. When the Gamma and Alpha indices are compared, it is true that northern Virginia scores higher on both, indicating that in terms of the number of edges or links and the number of circuits, the

**TABLE 4.7   Gamma and Alpha Indices for Road
Networks in Northern Virginia and Montgomery
County, Maryland**

	N. Virginia	Montgomery
No. of vertices	139	106
No. of edges	194	144
Gamma Index	0.472019	0.461538
Alpha Index	0.205128	0.188406

road network in northern Virginia is slightly better connected. But the differences between the two regions are quite small.

The Gamma and Alpha indices give us some general ideas about the overall connectivity of the two networks. It will be also interesting to see if different parts of the same network are equally accessible. In other words, we need to analyze the accessibility of the network locally, and the number of direct links for each road segment is a good indicator. Following the procedure for calculating the two indices, we choose the **Direct Links** menu item under the **Analysis** menu. This function calculates the direct links for each road segment. A larger number of links indicates better accessibility. After this menu item is selected, the number of direct links for each segment will be calculated and the result is stored as the new attribute link number in the attribute table. We conduct the same calculation for both regions. Because the numbers of direct links are now stored in the attribute table, these numbers can be used as an attribute for mapping. To do that, follow the steps below:

1. Double click the legend of the theme to invoke the **Legend Editor**.
2. In the **Legend Editor**, choose **Graduated Symbol** as the **Legend Type**.
3. In the **Classification Field**, choose **Link** as the field.

The width of a road segment is now proportional to the number of direct links of that road segment.

The results are shown in Figure 4.20. It is clear that road segments in both regions have many direct links. Closer visual inspection reveals that most segments in northern Virginia have multiple direct links, while certain road segments at the fringe of Montgomery County have relatively few direct links. Road segments with many direct links are limited to the so-called capital beltway and the Interstate 270 corridor. Road segments outside of these areas have fewer direct links. On the other hand, the direct links are quite evenly distributed in northern Virginia.

In our previous discussion, we included two other measures of accessibility: the number of steps required to reach the farthest edge of the network for each segment and the total number of direct and indirect links for each segment. These two measures are available by choosing the menu item **Steps** and **All Links** under

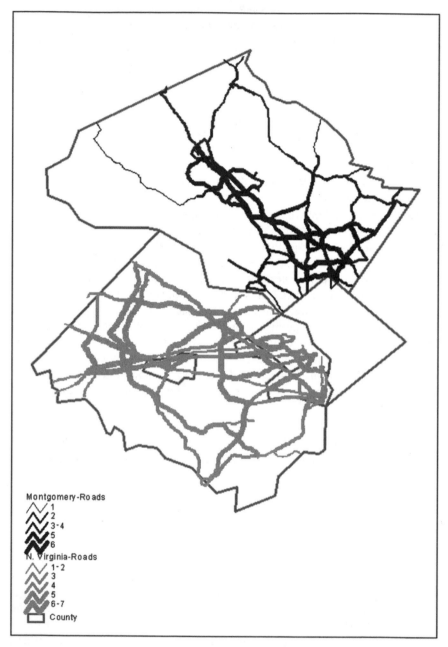

Figure 4.20   Number of direct links, selected northern Virginia and Montgomery County road networks.

the **Analysis** item. Repeating the set of steps we used before but choosing this menu item, the two measures for each road segment were added to their feature attribute tables. Then, modifying the legend via the **Legend Editor** and using **Graduate Symbol**, we created maps representing these two measures. They are shown in Figure 4.21 and 4.22.

In Figure 4.21, it seems that certain road segments in northern Virginia are less accessible than those in Montgomery County because those segments require 19 steps in order to reach the farthest edge of the network. This conclusion, however, is not valid because these two road systems have different sizes. The northern Vir-

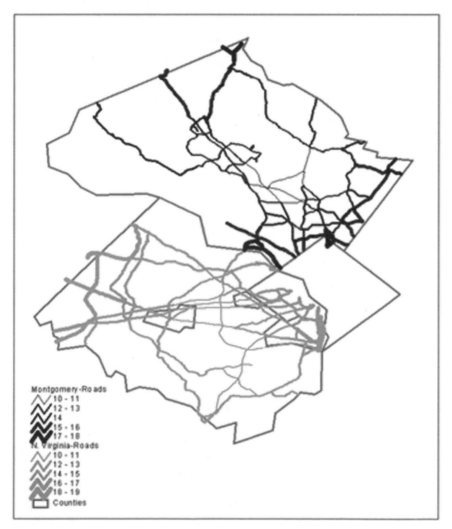

Figure 4.21 Numbers of steps reaching the entire network, selected northern Virginia and Montgomery County road networks.

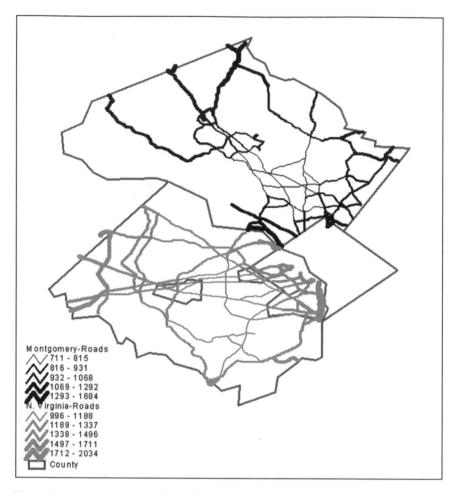

Figure 4.22 Total number of links (direct and indirect) in selected northern Virginia and Montgomery County road networks.

ginia system has more edges and nodes than the system in Montgomery County. Therefore, using this measure to compare the two areas is not appropriate. But within the same region, the number of steps is quite effective in indicating the relative accessibility of different locations. The central and southern parts of northern Virginia are highly accessible using this measure. In Montgomery County, the most accessible section is slightly north of the capital beltway in the southeastern section of the county.

Using the total number of direct and indirect links yields similar results. As mentioned before, we cannot compare the two regions on the basis of this measure because their networks have different sizes. On the basis of the total number of direct and indirect links alone, it seems that northern Virginia's system (highest 2034) is not very efficient because the number of links is much higher than that in Montgomery County (highest, 1,684), but northern Virginia's system is larger. Within the two regions, the results closely resemble those achieved using the number of steps. However, Figures 4.21 and 4.22 are very effective in showing the relative accessibility of road segments.

REFERENCES

DeMers, M. N. (2000). *Fundamentals of Geographic Information System.* (2nd edition). NY: John Wiley & Sons.

Taaffe, E. J., Gauthier, H. L., and O'Kelly, M. E. (1996). *Geography of Transportation.* (2nd edition). Upper Saddle River, NJ: Prentice-Hall.

Swan, A. R. H., and Sandilands, M. (1995). *Introduction to Geological Data Analysis.* Oxford: Blackwell Science.

CHAPTER 5

PATTERN DESCRIPTORS

Human settlements appeared in places where there were resources to support the population and where climate allowed such development. Changes in animal habitats often happened when events that altered their environment occurred. Locations of geographic objects form various spatial patterns according to their characteristics. Changes in spatial patterns over time illustrate spatial processes, as dictated by the underlying environmental or cultural factors.

The spatial patterns of geographic objects are often the result of physical or cultural processes taking place on the surface of the earth. *Spatial patterns* is a static concept since these patterns only show how geographic objects distribute at one given time. However, *spatial processes* is a dynamic concept because these processes show how the distribution of geographic objects changed over time. For any given geographic phenomenon, we often need to study both its spatial patterns and the spatial processes associated with these patterns.

Understanding the spatial patterns allows us to understand how the geographic phenomenon distributes and how it can be compared with others. The ability to describe spatial processes enables us to determine the underlying environmental or cultural factors that are changing the patterns. If the changes were desirable, we would find ways to promote them. If the changes were not desirable, we would need to find ways to correct the problems.

Spatial statistics are the most useful tools for describing and analyzing how various geographic objects (or events) occur or change across the study area. These statistics are formulated specifically to take into account the locational attributes of the geographic objects studied. We can use spatial statistics to describe the *spatial patterns* formed by a set of geographic objects so that we can compare them with patterns found in other study areas. For the spatial processes associated with these patterns, we can use spatial statistics to describe their forms, to detect changes, and to analyze how some spatial patterns change over time.

In earlier chapters, we demonstrated the use of descriptive statistics to measure central tendency and dispersion among point-based geographic objects. In this chapter, we will discuss the use of spatial statistics to describe and measure spatial patterns formed by geographic objects that are associated with *areas*, or *polygons*. We will use spatial statistics to describe spatial patterns exhibited in a set of polygons according to their characteristics. In addition, we will examine how the measured spatial patterns can be compared by using tests based on these spatial statistics.

5.1 SPATIAL RELATIONSHIPS

When studying a spatial pattern, we may want to compare that pattern with a theoretical pattern to determine if the theory holds in the case of the pattern being studied. Alternatively, we may want to classify spatial patterns into an existing categorical structure of known patterns. If the spatial patterns we study correspond to a particular theoretical pattern, we will be able to apply properties of the theory to interpret the spatial pattern we study. Or, if the spatial pattern can be closely related to a known pattern, we will be able to borrow from experience with and knowledge of the pattern for further study. In either case, it is necessary for us to establish a structure of categories of the spatial pattern.

A spatial pattern can be *clustered*, *dispersed*, or *random*. In Figure 5.1, these three possibilities are shown in hypothetical patterns in the seven counties of northeastern Ohio (Cuyahoga, Summit, Portage, Lake, Geauga, Trumbull, and Ashtabula). In Case 1, the darker shade representing certain characteristics associated with the counties appear to have *clustered* on the western side of the seven-county area. On the eastern side, the lighter shade (the other type of characteristics) prevails among the remaining counties. Perhaps the darker shade indicates the growth of the urban population in recent years, whereas the lighter shade indicates the loss of the urban population during the same period.

In Case 2, counties that are shaded darker appear to be spaced evenly and away from each other. This is often referred to as a *dispersed* pattern. It suggests that the geographic phenomenon displayed may be uniformly distributed across this seven-county area. For example, it may be that the political preferences between the two parties in these counties happen to change between neighboring counties. For an extremely uniform or dispersed pattern, the checkerboard in which black and white cells distribute systematically and change between every pair of neighboring cells is a typical example of a dispersed spatial pattern. A repulsive spatial relationship among neighbors is implied.

The spatial pattern in Case 3 of Figure 5.1 does not appear to be either clustered or dispersed. It may be close to what we typically call the *random* pattern. If a spatial pattern is random, it suggests that there may not be any particular sys-

Case 1: Clustered

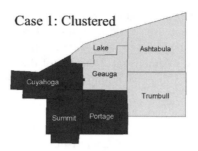

Case 2: Dispersed

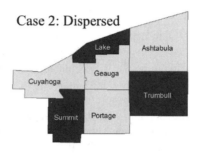

Case 3: Random

Figure 5.1 Types of patterns: clustered, dispersed, and random.

tematic structure or any particular mechanism controlling the way these polygons form the pattern.

In most geographic problems, there is no clear indication that the pattern is clustered, dispersed, or random. Rather, most real-world patterns are somewhere between a random pattern and a dispersed pattern or between a random pattern and a clustered pattern. Very rarely would we find a spatial pattern that is extremely clustered, extremely dispersed, or purely random.

Since real-world patterns can rarely be clearly classified random, clustered, or dispersed, the question is, how close is a given spatial pattern to any of the three patterns? In the next section, we will discuss how we measure a spatial pattern to determine its closeness to a random pattern so that the pattern can be further studied.

5.2   SPATIAL AUTOCORRELATION

In classifying spatial patterns as either clustered, dispersed, or random, we can focus on how various polygons are arranged. We can measure the similarity or dissimilarity of any pair of neighboring polygons. When these similarities and dissimilarities are summarized for the spatial pattern, we have spatial autocorrelation (Odland, 1988).

Spatial autocorrelation means that the attribute values being studied are self-correlated and the correlation is attributable to the geographic ordering of the objects.

Many situations exhibit some degree of spatial autocorrelation. When comparing agricultural production levels among all the farms in a region, we would not find that all farms are producing at the same rate. Even though the local climate may be the same for all of these farms, the soil conditions or water supplies within the region may vary. Still, neighboring farms within the region share similar soil and moisture conditions, so these farms probably have similar levels of production.

Among population densities measured in each county, we would see that those counties located in or close to big central cities tend to have higher population densities than those located far away. This is because the outward influence of the central city often makes the surrounding counties its hinterlands. Once the underlying reason is identified, it would be logical to investigate why urban population densities cluster around big central cities or why the production levels of farms cluster or disperse along known soil structures or water supplies.

The basic property of spatially autocorrelated data is that the values are not random in space. Instead, the data values are spatially related to each other, even though they may be related in different ways. Referring back to the three cases in Figure 5.1, Case 1 has a *positive* spatial autocorrelation, with adjacent or nearby polygons having similar shades (values). In Case 2, the dispersed pattern has a *negative* spatial autocorrelation, with changes in shade often occurring between adjacent polygons. As for Case 3, it appears to be close to a random pattern in which little or no spatial autocorrelation exists.

In addition to its type or nature, spatial autocorrelation can be measured for its strength. Strong spatial autocorrelation means that the attribute values of adjacent geographic objects are strongly related (either positively or negatively). If the attribute values of adjacent geographic objects appear to have no clear order or relationship, the distribution is said to have a weak spatial autocorrelation or a random pattern. Figure 5.2 shows five different structures with darker and lighter shades within the seven counties in northeastern Ohio. Opposite to the numeric line, the patterns toward the left end are examples of positive spatial autocorrelation, while the patterns toward the right end are examples of negative spatial

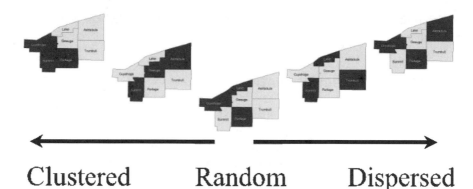

Clustered   Random   Dispersed

Figure 5.2   Spatial autocorrelation.

autocorrelation. In the middle is a pattern of weak spatial autocorrelation or close to a random pattern.

Spatial autocorrelation can be a valuable tool to study how spatial patterns change over time. Results of this type of analysis often lead to further understanding of how spatial patterns change from the past to the present, or to estimations of how spatial patterns will change from the present to the future. Therefore, useful conclusions can be drawn to advance our understanding of the underlying factors that drive the changes in spatial patterns.

In addition, the study of spatial autocorrelation has significant implications for the use of statistical techniques in analyzing spatial data. For many classical statistics including various regression models, a fundamental assumption is that observations are randomly selected and independent of each other. But when spatial data are analyzed, this assumption of independence is often violated because most spatial data have certain degrees of spatial autocorrelation (Anselin and Griffith, 1988). If spatial autocorrelation in the data is significant but disregarded, the results will be incorrect. For example, the regression parameters in a regression model that is constructed using classical statistics on geographic data may not be biased, but the significant testing of the parameters could yield wrong conclusions. Consequently, measuring and testing the significance of spatial autocorrelation is essential before any statistical analysis is conducted.

To measure spatial autocorrelation, there are statistics that allow us to work with points or polygons. Since the methods for measuring spatial autocorrelation in point patterns have been discussed in Chapter 3, we will focus on the methods suitable for polygons. These methods may be used to measure spatial autocorrelation of nominal and interval/ratio data. Specifically, joint count statistics can be used to measure spatial autocorrelation among polygons with binary nominal data. For interval or ratio data, we will discuss Moran's I Index, the Geary Ratio, Local Indicators for Spatial Association, and the G-statistics. In addition, a graphical tool, the Moran scatterplot, used to present spatial autocorrelation visually, is included.

5.3 SPATIAL WEIGHTS MATRICES

For measuring spatial autocorrelation in a set of geographic objects, we have to discuss methods for capturing spatial relationships among areal units, or polygons. Recall that spatial autocorrelation measures the degree of sameness of attribute values among areal units within their neighborhood. The concept of *neighborhood* has to be quantified so that it can be applied in the calculation of spatial autocorrelation statistics. In other words, the neighborhood relationship among areal units has to be captured before we can calculate the statistics.

Assume that we have n areal units in our study area. Given any predefined method to determine the neighborhood relation for these n areal units, we have $n \times n$ pairs of relationship to be captured. Conventionally, we use a matrix (in this case, $n \times n$ in dimension) to store and organize the spatial relationship among these n areal units. Each areal unit is represented by a row and a column. Each value in the matrix indicates the spatial relationship between the geographic features represented by the corresponding row and column. However, given different criteria to define the neighborhood relationship, we may derive different matrices (Griffith, 1996). For example, a binary value of 1 or 0 can be used to represent whether or not two polygons are spatially adjacent. Alternatively, actual measures of distances between centroids of polygons can be stored in such a matrix to represent their spatial adjacency in a different way.

In the following subsections, we discuss different ways to specify spatial relationships and their associated matrices. In general, these matrices are called *spatial weights matrices* because elements in the matrices are often used as weights in the calculation of spatial autocorrelation statistics.

5.3.1 Neighborhood Definitions

There are many ways to define a spatial relationship. Even if we are concerned with the immediate neighbors of an areal unit, there are at least two common methods. Figure 5.3 illustrates the *rook's case* and the *queen's case* of neighbor relationships in a grid. In a highly simplified polygon structure similar to a set of grid cells, there are nine areal units, with the one we are concerned about at the center (X). Using the rook's case as the criterion to define a neighbor, only

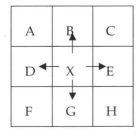 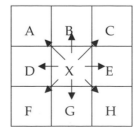

Figure 5.3   Neighbor definitions: rook's and queen's cases.

B, D, E, and G are neighbors because each of them shares a boundary with the polygon X. If we adopt the queen's case, then all the surrounding areal units can be identified as neighbors of X as long as they touch each other even at a point.

The neighborhood definitions described in Figure 5.3 correspond to the *adjacency measure*. If any two areal units are next to each other (juxtaposed), then they are neighbors of each other. In the rook's case, the neighboring units have to share a boundary with length greater than 0. In the queen's case, the shared boundary can be just a point along the diagonal polygons. The neighboring polygons of X identified by these two criteria are its immediate or first order neighbors. The adjacency measure can be extended to identify neighbors of the immediate neighbors or, more technically, second order neighbors. The concept can be further extended to identify higher order neighbors, even though computationally it may sometimes not be trackable.

Besides adjacency as the criterion used to define a neighborhood, another common measure is *distance*. Distance is a common measure of a spatial relationship. Corresponding to the geographic concept of *distance decay*, intensity of spatial activities between two distant places are said to be less than the intensity of activities between places that are closer together, assuming that all places have the same characteristics. Consequently, the distance between any two geographic features can be used as an indicator of their spatial relationship. The distance measure can also be used to define a neighborhood.

If, for a household, neighbors means those houses are less than 1 mile, then its neighborhood can be quantitatively defined as all houses within a radius of 1 mile from the household. With this definition, we can check all houses to see if they fall within this area. The results can be displayed in a binary form (inside or outside of the defined neighborhood).

5.3.2 Binary Connectivity Matrix

With different ways of defining neighbors, different matrices can be constructed to capture the spatial relationship among geographic features. Using the simplest adjacency definition of neighborhood, areal units sharing a common boundary (at this point, we put aside the difference between the rook's and queen's cases) are neighbors. In the matrix capturing the spatial relationship, we can put a 1 in the cell corresponding to the two geographic features or polygons if they are next to each other. If any two polygons are not adjacent to each other, the corresponding cell will have a value of 0. In the entire matrix, the cell value is either 0 or 1. This type of matrix is sometime called a *binary matrix*. Because this matrix also shows how pairs of polygons are connected, and because a similar concept can be applied to linear features in a network (Chapter 4), this binary matrix is also called a *connectivity matrix*. In such a matrix, C, c_{ij} denotes the value (either 0 or 1) in the matrix in the ith row and the jth column.

A binary connectivity matrix has several interesting characteristics. First, all elements along the major diagonal, c_{ij} (the diagonal from upper left to lower right), are 0 because it is assumed that an areal unit is not a neighbor of itself.

Second, this matrix is symmetric, i.e., the upper triangle of the matrix divided along the major diagonal is a mirror of the lower triangle. Using our notation, $c_{ij} = c_{ji}$. This symmetric property of the matrix basically reflects the reciprocal nature of spatial relationship: areal unit A is a neighbor of B, and B is also a neighbor of A. The symmetric structure of the matrix requires information on the spatial relationship to be stored twice—a redundancy in this type of matrix. Finally, a row in the matrix represents how one areal unit is related spatially to all other units. Therefore, summing the cell values for that row across all columns, also known as the *row sum*, indicates the number of neighbors that the areal unit has. Using our notation, the *row sum* is

$$c_{i.} = \sum_j c_{ij}.$$

Table 5.1 is a binary connectivity matrix of the seven Ohio counties using the queen's case. Therefore, Cuyahoga County and Portage County are neighbors. Similarly, Geauga and Summit counties are depicted as neighbors in the matrix. The sum of all these cell values (26) is twice the number of shared boundaries or joints ($2J$).

Apparently, the binary matrix is not a very efficient format to store spatial relationship based upon adjacency. Not only the upper triangle of the matrix duplicates the lower triangle; in most spatial configurations, most of the cell values in the matrix are 0s, indicating that the two areal units are not neighbors. In the example of the seven Ohio counties, this characteristic of the matrix may not be obvious. But still, among these seven areal units with a matrix of 49 cells, there are 23 0s. Imagine creating a binary matrix for the approximately 3,000 counties of the entire United States; the matrix would not only be gigantic ($3,000 \times 3,000 - 9$ million cells), but most of the cell values would be 0s. Most of the time, we are only interested in neighbors. Using a lot of space to store nonneighbor information is not efficient.

Instead of using a square matrix ($n \times n$) to record areal units that are neighbors and nonneighbors, a more compact format of the matrix includes only areal units that are neighbors of a given unit. In general, each areal unit has a unique identification number, such as the Federal Information Processing Standards (FIPS) code

TABLE 5.1 Binary Connectivity Matrix (Seven Ohio Counties)

ID	Geauga	Cuyahoga	Trumbull	Summit	Portage	Ashtabula	Lake
Geauga	0	1	1	1	1	1	1
Cuyahoga	1	0	0	1	1	0	1
Trumbull	1	0	0	0	1	1	0
Summit	1	1	0	0	1	0	0
Portage	1	1	1	1	0	0	0
Ashtabula	1	0	1	0	0	0	1
Lake	1	1	0	0	0	1	0

TABLE 5.2   A Sparse Matrix Capturing the Same Information as the Binary Matrix

ID	Neighbor 1	Neighbor 2	Neighbor 3	Neighbor 4	Neighbor 5	Neighbor 6
Geauga	Cuyahoga	Trumbull	Summit	Portage	Ashtabula	Lake
Cuyahoga	Geauga	Summit	Portage	Lake		
Trumbull	Geauga	Portage	Ashtabula			
Summit	Geauga	Cuyahoga	Portage			
Portage	Geauga	Cuyahoga	Trumbull	Summit		
Ashtabula	Geauga	Trumbull	Lake			
Lake	Geauga	Cuyahoga	Ashtabula			

of a county (which combines the state and county FIPS codes), or a unique label such as the county name (unique only within the state). Using the unique identifier, a rather compact (sparse) matrix listing only the identifiers of neighbors can be constructed. Table 5.2 captures the same information as Table 5.1, but its size is greatly reduced. Instead of 49 cells (7×7) with many 0s, the dimension of the matrix is now 7×6, with only 26 cells containing information. In a much larger area with more polygons, the reduction in matrix size will be more dramatic.

5.3.3   Stochastic or Row Standardized Weights Matrix

The binary matrix basically has a weight of 1 if the areal unit is a neighbor. Mathematically, this unit weight may not be very effective in modeling the spatial relationship. For instance, we want to analyze how a house value is affected by the values of its surrounding units. Following the normal practice of realtors, we can think of the house value as receiving a fractional influence from each of its neighbors. If there are four neighbors for that house, its value may receive 0.25 influence from each neighboring house.

Recall that the binary matrix consists of 1s and 0s. A 1 indicates that the corresponding areal units represented by the rows and the columns are neighbors. Therefore, for each given row, the row sum, $c_{i.}$, indicates the total number of neighbors that areal unit has. To find out how much each neighbor contributes to the value of the areal unit of concern when each neighbor exerts the same amount of influence on that areal unit, we would calculate the ratios of each house with respect to the total influence. This gives the weight (w_{ij}) of each neighboring unit:

$$w_{ij} = c_{ij}/c_i.$$

Figure 5.4 uses the seven Ohio counties to illustrate how the *row-standardized matrix* or the *stochastic matrix*, which is usually denoted as W, can be derived from the binary matrix C. Please note that even the matrix still has a major diagonal of 0s; this matrix is no longer symmetric.

Binary Connectivity Matrix

ID	Geauga	Cuyahoga	Trumbull	Summit	Portage	Ashtabula	Lake	Row Sum
Geauga	0	1	1	1	1	1	1	6
Cuyahoga	1	0	0	1	1	0	1	4
Trumbull	1	0	0	0	1	1	0	3
Summit	1	1	0	0	1	0	0	3
Portage	1	1	1	1	0	0	0	4
Ashtabula	1	0	1	0	0	0	1	3
Lake	1	1	0	0	0	1	0	3

Stochastic or Row-Standardized Matrix

ID	Geauga	Cuyahoga	Trumbull	Summit	Portage	Ashtabula	Lake
Geauga	0	0.17	0.17	0.17	0.17	0.17	0.17
Cuyahoga	0.25	0	0.00	0.25	0.25	0.00	0.25
Trumbull	0.33	0.00	0	0.00	0.33	0.33	0.00
Summit	0.33	0.33	0.00	0	0.25	0.00	0.00
Portage	0.25	0.25	0.25	0.25	0	0.00	0.00
Ashtabula	0.33	0.00	0.33	0.00	0.00	0	0.33
Lake	0.33	0.33	0.00	0.00	0.00	0.33	0

Figure 5.4 Deriving the stochastic weight matrix from the binary connectivity matrix.

5.3.4 Centroid Distances

Besides using adjacency as a measure to describe the spatial relationship among a set of geographic features and to define a neighborhood among them, another common measure is *distance*. Using distance as the weight in describing a spatial relationship is very powerful. Recall the first law of geography summarized by Waldo Tobler (1970): all things are related, but closer things are more related. In other words, the relationship between any two geographic features is a function of distance between them. Generally, we expect features close to each other to be related. Therefore, using distance as the weight to depict a spatial relationship is theoretically sound.

The distance between two point features is easy to define. There are several ways to measure the distance between any two polygons. A very popular method, especially in transportation studies, is to use the centroid of the polygon to represent that polygon. The centroid is the geometric center of a polygon. There are different ways of determining the centroid of a polygon. Each of them identifies centroids differently. But in general, the shape of a polygon affects the location of its centroid. Polygons with unusual (geometrically irregular) shapes may generate centroids located in undesirable locations. For instance, certain methods for centroid calculation may generate a centroid of the state of Florida located in the Gulf of Mexico, or the centroid of the state of Louisiana to be in the neighboring state of Mississippi. Still, in most cases and with the advances in algorithms in determining centroids, it is quite common, using the distance between the two centroids of the two polygons, to represent the distance between the polygons themselves.

Because distance is used as the weight, the spatial weights matrix is sometimes labeled D and the elements are labeled d_{ij}, representing the distance between (the centroids of) areal units i and j. Table 5.3 shows the D matrix using centroid distances for the seven Ohio counties. In modeling spatial processes, the distance weight is often used in an inverse manner, as the strengths of most spatial relationships diminish when distance increases. Therefore, when the distance matrix is used, the weight

$$w_{ij} = \frac{1}{d_{ij}}$$

TABLE 5.3 **Spatial Weights Matrix Using Centroid Distance (Seven Ohio Counties)**

ID	Geauga	Cuyahoga	Trumbull	Summit	Portage	Ashtabula	Lake
Geauga	0	25.1508	26.7057	32.7509	25.0289	26.5899	12.6265
Cuyahoga	25.1508	0	47.8151	23.4834	31.6155	50.8064	28.2214
Trumbull	26.7057	47.8151	0	41.8561	24.4759	29.5633	36.7375
Summit	32.7509	23.4834	41.8561	0	17.8031	58.0869	42.7375
Portage	25.0389	31.6155	24.4759	17.8031	0	47.5341	37.4962
Ashtabula	26.5899	50.8064	28.5633	58.0869	45.5341	0	24.7490
Lake	12.6265	28.2214	36.7535	42.7375	37.4962	24.7490	0

is an inverse of the distance between features i and j. In other words, the weight is inversely proportional to the distance between the two features. Based upon empirical studies of spatial processes, however, the strength of many spatial relationships has been found to diminish more than proportionally to the distance separating the features. Therefore, the squared distance is sometimes used in the following format:

$$w_{ij} = \frac{1}{d_{ij}^2}.$$

These weighting schemes of inverse distance will be used later for different measures of spatial autocorrelation.

5.3.5 Nearest Distances

With the advances in GIS software algorithms, features other than the distance between centroids can be easily determined. It is also relatively easy to determine the distance between any two geographic features based on the distance of their nearest parts. Selecting the two farthest counties among the seven Ohio counties, the nearest parts between Summit and Ashtabula counties are the northeastern corner of Summit and the southwestern corner of Ashtabula (Figure 5.1). This conceptualization of distance between geographic features is potentially useful if the investigator is interested in spatial contact or diffusion.

An interesting situation involving the distance of nearest parts is when the two features are adjacent to each other. When this is the case, the distance between two neighboring features is 0. In other words, under this distance measurement scheme, 0 distance means that the corresponding features are immediate neighbors. The 1s in a binary connectivity matrix also capture this information. By extracting all the nondiagonal 0 cells from the distance matrix, we can also derive the binary matrix. Since the binary connectivity matrix can be seen as a simplified distance weight matrix, it is sometimes said that the binary matrix is a derivative of the distance matrix based on nearest parts. Please note that the above discus-

TABLE 5.4   Spatial Weights Matrix Based upon Nearest Parts (Seven Ohio Counties)

ID	Geauga	Cuyahoga	Trumbull	Summit	Portage	Ashtabula	Lake
Geauga	0	0.0000	0.0000	0.0000	0.0000	0.0000	0.0000
Cuyahoga	0.0000	0	0.3561	0.0000	0.0000	0.3614	0.0000
Trumbull	0.0000	0.3561	0	0.3705	0.0000	0.0000	0.1670
Summit	0.0000	0.0000	0.3705	0	0.0000	0.4015	0.2179
Portage	0.0000	0.0000	0.0000	0.0000	0	0.1518	0.2180
Ashtabula	0.0000	0.3614	0.0000	0.4015	0.1518	0	0.0000
Lake	0.0000	0.0000	0.1670	0.2179	0.2180	0.0000	0

sions on inverse distance weighting schemes are also applicable to the current distance measure.

Table 5.4 is the distance matrix based on nearest parts of the seven Ohio counties. Compare this matrix with the binary matrix in Table 5.1. The nondiagonal cells in the distance matrix with 0 values correspond to the cells with 1s in the binary matrix.

ArcView Notes As in previous chapters, a project file was designed for this chapter (Ch5.apr). It is located on the companion website to this book.

In addition to the standard ArcView menu categories, a new menu category, **Spatial Autocorrelation**, has been added. Under this menu category, the first menu item is **Creating Weights Matrices**. Please be sure that a View window with the interested theme is active (i.e., clicked on). If multiple themes were added to the View window, the procedure assumes that the *first theme* is the one for which the analyst is interested in performing spatial statistical analysis.

When this menu item is selected, the procedure will first check to see if any feature on the **View** window has been selected. If no feature has been selected, the procedure will select all features appearing in the View window to construct the spatial weights matrix. In other words, if users prefer to create a weights matrix for a subset of features on the View window, the user should either

1. use the selection tool or **QueryBuilder** in ArcView to select the features desired before choosing the menu item to create the matrix or
2. choose **No** when the procedure asks users to select all features when users forgot to select features before proceeding to weights matrix creation.

After selecting the desired features, run the weights matrix procedure again. If only a subset of all features is selected to create the weights matrix, users should remember, in subsequent analyses, to use only the subset of features instead of all features in the View window.

With the features for calculation selected, users will be asked to select a field in the table as the *identification (ID) field*. Please note that this ID field has to provide a unique identifier for each feature included in the analysis. In the seven Ohio counties example, the combined state or county FIPS code would be appropriate because no two counties in the United States have the same combined state and county FIPS code. We can even use county names as the ID field in this example because we know that each of the seven counties has a unique name. In a land use analysis, however, using types of land use as the ID field may be inappropriate because they are probably multiple polygons or areal units with the same land use type.

The procedure will provide users with four choices for weights matrices: `binary connectivity`, `stochastic weight`, `distance between nearest parts`, and `centroid distance`. Choose one of these options; the procedure will ask for the name of the output file.

Please note that the procedures for weights matrix generation developed here are meant for small data sets. They are not intended to support major research or analysis with large numbers of polygons or large spatial systems.

The computation speed also varies, depending on the type of matrix selected (in addition to other factors, such as hardware configurations) and the speed of the hardware. In general, the procedure is quite fast for the centroid distance matrix. For the matrices on binary connectivity and the distance for nearest parts, the procedure can be completed in a reasonable time frame (less than 1 minute on an average desktop computer) for spatial systems as large as 50 areal units. For larger spatial systems, the time required is not proportional but is exponentially related to the increase in the number of areal units.

The procedure is slowest in constructing the stochastic weights matrix because it requires additional steps to derive the total number of neighbors and then to perform the standardization. Still, these steps add just a few more seconds for a spatial system with 50 areal units. For larger systems, the extra time may be longer. But overall, this spatial weights matrix procedure handles small spatial systems very well.

5.4 TYPES OF SPATIAL AUTOCORRELATION MEASURES AND SOME NOTATIONS

In this section, we will discuss different methods for calculating spatial autocorrelation statistics utilizing one or more of the spatial weights matrices discussed above. Different statistics are used to handle attributes that are measured at different scales. In addition, different statistics capture different aspects of spatial autocorrelation. In this section, several statistics are introduced.

If the spatial attributes or variables to be studied are measured in *nominal scale* and are *binary* (i.e., the attribute has only two possible values), then *joint count statistics* can be used. If the spatial variables are measured in *interval* or *ratio* scale, the appropriate spatial autocorrelation statistics are *Moran's I* index and *Geary's C Ratio*. Another possible choice is the general *G-statistic*.

All of these measures can be regarded as *global measures of spatial autocorrelation* or *spatial association* because one statistic or value is derived for the entire study area, describing the overall spatial relationship of all the areal units. However, there is no reason to believe that any spatial process is homogeneous within the distribution itself. The magnitude of spatial autocorrelation can vary by locations, and thus a distribution or a spatial pattern can be spatially heterogeneous. To describe the spatial heterogeneity of spatial autocorrelation, we have to rely on measures that can detect spatial autocorrelation at a local scale. The *Local Indicator of Spatial Association (LISA)* and the *local G-statistics* are designed for this purpose.

In deriving various spatial autocorrelation statistics and related statistics to test for their significance, several terms derived from the spatial weights matrices will be used repeatedly. Therefore, it is logical to introduce and define them here before we discuss those spatial autocorrelation statistics.

Even though a weight, w_{ij}, is often used to represent the cell value of the stochastic weights matrix, W, for row i and column j, it is also quite common to use w_{ij} to represent the cell value of any weights matrix. For any given weights matrix, summing up all cell values of a given row i across all columns (row sum) is denoted as

$$w_{i.} = \sum_j w_{ij}.$$

Similarly, a column sum is the sum for a given column j across all rows:

$$w_{.j} = \sum_i w_{ij}.$$

Sometimes W also represents the sum of all cell values of the weights matrix:

$$W = \sum_i \sum_j w_{ij}.$$

In testing the significance of several spatial autocorrelation statistics, the weight structure has to be summarized by several parameters, including SUM_1 and SUM_2. These two terms are defined as

$$SUM_1 = \frac{1}{2} \sum_i \sum_j (w_{ij} + w_{ji})^2$$

and

$$SUM_2 = \sum_i \left(\sum_j w_{ij} + \sum_j w_{ji} \right)^2.$$

The first term, SUM_1, is the sum over the weights. If the weights are binary and the matrix is symmetric (i.e., the C matrix), then $(w_{ij} + w_{ji})^2 = 4$. The SUM_1 is simply four times the total number of joints or shared boundaries in the entire study area. The second term, SUM_2, is based upon the sums of the weights associated with each areal unit first, but in both directions (i.e., for both w_{ij} and w_{ji}). The sums are then added, squared, and summed over all areal units.

Let's use n to denote the number of areal units in the entire study area. If there are two groups of areal units, for instance, defined by an attribute, which carries two values, x and y, conventionally we use n_x and n_y to indicate the number of areal units in the two groups. Similar to the notation but very different in meaning, we can use

$$n^{(x)} = n * (n - 1) * (n - 2) * (n - 3) * \cdots * (n - x + 1)$$

where $n > x$. For example, if $n = 5$, $n^{(3)} = n(n - 1)(n - 2) = 5 \times 4 \times 3$ and $n^{(1)} = n$.

If x_i is the attribute value for areal unit i, a new parameter, m_j, can be derived based upon x_i, as

$$m_j = \sum_{i=1} x_i^j,$$

where $j = 1, 2, 3, 4$. Therefore, if $j = 1$, m_j is the sum of x_i of all i. If $j = 2$, m_j is the sum of all the squares of x_i.

All these terms will be used later in this chapter for the discussions of various spatial autocorrelation measures. Readers should refer back to this section later as needed. For the rest of this chapter, we will discuss different spatial autocorrelation measures, starting with joint count statistics for binary data and proceeding to global measures (including Moran's I, the Geary Ratio, and the general-G statistic) and local measures (including LISA and local G-statistics) for interval and ratio data. Finally, we will discuss the Moran scatterplot, a graphical technique used to visualize the spatial heterogeneity of spatial autocorrelation.

5.5 JOINT COUNT STATISTICS

The use of *joint count statistics* provides a simple and quick way of quantitatively measuring the degree of clustering or dispersion among a set of spatially adjacent polygons. This method is applicable to *nominal* data only. Because the statistics are based on comparing the actual and expected counts of various types of joints between adjacent polygons having the same or different attribute values, the nominal data appropriate for this method are limited to *binary* data. Binary data are those with only two possibilities, such as arable/nonarable lands, high/low income groups, urban/rural counties, and so on. To simplify our description, let's use black and white to indicate the two possible attribute values associated with polygons. Consequently, the various types of joints would be black-black joints, black-white joints, and white-white joints.

If a polygon pattern has a clustered pattern, such as that of the Case 1 in Figure 5.1, we would expect to have more white-white or black-black joints than white-black joints. This situation is also known as having *positive spatial autocorrelation*, or a pattern in which similar values are close to each other. Alternatively, a dispersed pattern, or *negative spatial autocorrelation*, has more black-white joints than black-black or white-white joints. Case 2 in Figure 5.1 is an example of a dispersed pattern. Of course, for a random pattern, we would expect the actual counts of various types of joints to be fairly close to those of a typical random pattern.

While the joint count statistics method is limited to binary data, data in other measurement scales can be easily downgraded or converted to binary form. For a set of ordinal data values, one can set a rank as the cutoff level so that those values above this level are counted as one type and those below it are counted as another type. In ranking cities by their population sizes, those cities ranked larger than Cleveland are assigned one value and those cities ranked smaller than Cleveland are assigned another value. Similarly, downgrading interval/ratio data is only a matter of setting a cutoff point to separate the values above and below this level into two binary values. Take average family income as an example. One can consider cities with an average family income above the national average as one type and cities with an average family income below the national average as another type. Therefore, the two essential conditions for using the joint count statistics are as follows:

1. Data are related to area (polygons).
2. Data are measured in binary form (only two possible values exist).

Please note that although it is feasible to convert data from interval/ratio and ordinal scales into binary form, this process also reduces the amount of information captured by the data, so we normally try to avoid this process if possible. Especially if interval or ratio data are converted into nominal scale, the precision of the original data is lost.

The joint count statistics method calculates the difference in the number of black-black, white-white, and black-white joints between the patterns to be tested. With small spatial systems, it is probably feasible to manually count the number of the three different types of joints in order to derive the statistics. But if the data are already in digital formats and the spatial systems are too large for manual operations, the counting processes have to be automated. Below are the general steps used to derive the three joint count statistics in a computational environment.

1. Let $x_i = 1$ if polygon i is a black polygon and $x_i = 0$ if it is white.
2. Then, for black-black joints, $O_{BB} = \frac{1}{2} \sum_i \sum_j (w_{ij} x_i x_j)$.
3. For white-white joints, $O_{WW} = \frac{1}{2} \sum_i \sum_j [w_{ij}(1 - x_i)(1 - x_j)]$.
4. For black-white or white-black joints, $O_{BW} = \frac{1}{2} \sum_i \sum_j [w_{ij}(x_i - x_j)^2]$.

Please note that the weight, w_{ij}, can be the binary weight or the row-standardized weight. The three statistics above are the observed joint counts describing the actual pattern. If we observe a large number of O_{BB} or O_{WW} joints, or both, we may postulate that the observed pattern may exhibit positive spatial autocorrelation or clustering. However, we cannot conclude that positive spatial autocorrelation exists until we demonstrate that the observed pattern is different from a random pattern and that the difference is probably not due to chance or coincidence. That is the concept of *likelihood.*

The user of the joint count statistics method, however, needs to know how to estimate the likelihood that each polygon has a white or a black value (the attribute value). Different ways of estimating attribute values for the polygons will affect the outcome of the joint count statistics process.

If the probability of a polygon's being black or white is based on known theories or a trend derived from a larger region, the method by which the attribute values are estimated is known as *free sampling.* This means that the probability of a polygon's being white or black is not limited or affected by the total number of black or white polygons in the group. Consequently, this approach is sometimes referred to as *normality sampling.* Alternatively, if the probability of a polygon's being black or white is limited by or dependent upon the total number of black or white polygons, the method by which the attribute values are estimated is known as *nonfree sampling* or *randomization sampling.*

In our seven-county example, the nonfree sampling case can only have three black polygons and four white polygons, no matter how they are rearranged. Since the total number of black and white polygons is fixed, the method is nonfree sampling, or *sampling without replacement.* Compared to nonfree sampling, free sampling does not limit how many polygons can be black or white; therefore, the method is also known as *sampling with replacement.*

When using joint count statistics, the choice between normality and randomization sampling is important. As a rule of thumb, the normality sampling approach should not be used whenever references to trends from larger regions or those outside of the study area cannot be used with certainty. This is because randomization sampling requires less rigorous assumptions than free sampling. Normality sampling should be used if the relationship between the study area and the national trend or the trend from a larger region can be established with known theories or by experience.

5.5.1 Normality Sampling

In both normality sampling and randomization sampling, calculating joint count statistics involves estimation of the expected number of black-black, white-white, and black-white joints and their standard deviations. The expected numbers of these joints reflect a random pattern, or a pattern with no significant spatial autocorrelation of any type. The number of black-black and white-white joints indicates the magnitude of positive spatial autocorrelation, while the number of black-white or white-black joints indicates the magnitude of negative spatial autocorrelation.

These observed values are compared with their expected counterparts to derive their differences. These differences are then standardized by their corresponding standard deviations in order to obtain standardized scores. Using these scores, we can decide if there is a significant positive or negative spatial autocorrelation in the pattern. In other words, three pairs of comparisons have to be conducted. For illustrative purpose, we will only show in detail how negative spatial autocorrelation can be tested. With this example, the other two situations can be repeated easily or derived using the accompanying ArcView project file.

For normality sampling, the equations for the expected number of black-black and white-white joints are

$$E_{BB} = \frac{1}{2} W p^2$$

and

$$E_{WW} = \frac{1}{2} W q^2.$$

Then the equation for expected black-white joints is

$$E_{BW} = W p q,$$

where

E_{BB}, E_{WW}, and E_{BW}, are the expected number of black-black, white-white, and black-white joints, respectively,

p is the probability that an area will be black, and

q is the probability that an area will be white.

The two probabilities must sum to 100%, or ($p+q = 1.0$). If no other information is available, a common method is to set $p = n_B/n$. But there are other considerations, to be discussed later, in determining p. If the spatial weights matrix is a binary matrix, the expected values can be simplified to:

$$E_{BB} = J p^2$$
$$E_{WW} = J q^2$$
$$E_{BW} = 2 J p q,$$

where J is the total number of joints in the study area.

To test if the measured pattern is significant enough, a statistical test, known as the *Z-test*, can be applied. To perform this test, the standard deviations of the expected joints are also needed. When a stochastic weights matrix is used, the three standard deviations are

$$\sigma_{BB} = \sqrt{\frac{1}{4} p^2 q [S_1 q + S_2 p]}$$

$$\sigma_{WW} = \sqrt{\frac{1}{4} q^2 p [S_1 p + S_2 q]}$$

$$\sigma_{BW} = \sqrt{\frac{1}{4} \{ 4 S_1 pq + S_2 pq * [1 - 4pq] \}}.$$

If a binary matrix is used, these formulas are reduced to

$$\sigma_{BB} = \sqrt{p^2 J + p^3 K - p^4 (J + K)}$$

$$\sigma_{BB} = \sqrt{q^2 J + q^3 K - q^4 (J + K)}$$

$$\sigma_{BW} = \sqrt{2pq J + pq K - 4p^2 q^2 (J + K)},$$

where σ are the standard deviations of the corresponding types of joints and J, p, and q are as defined previously. K is $\sum_{i=1}^{n} L_i (L_i - 1)$. The n in $\sum_{i=1}^{n} L_i (L_i - 1)$ is the total number of polygons, and L_i is the number of joints between polygon i and all the polygons adjacent to it.

Let us focus on testing negative spatial autocorrelation (i.e., black-white joints) using the binary matrix (and its corresponding formulas). When calculating the expected black-white joints, the probability of any polygon's being black or white must be known, namely, p and q. If this probability is known, the only step needed to complete the task is to count the number of joints, J.

Calculating J is quite straightforward since it only involves counting the number of joints between the polygons or summing up all values of the binary connectivity matrix and then dividing this number in half. For the example in Figure 5.5, the number of joints is 11. Consequently, we will have $J = 11$. Please note that all joints are to be counted for J, including all black-black, black-white, and white-white joints.

Two or more polygons may be contiguous at only one point rather than sharing borders. In this case, it will be up to the analyst to decide if this point should be counted as a joint. That is, the analyst must decide if the rook's case or the queen's case should be used in constructing the spatial weights matrix. If one joint of this type is included, it is imperative that all such joints are counted, both the total number of joints and the number of black-white joints. In this illustrative example, we adopt the rook's case in defining neighbors. Therefore, Cuyahoga County and Portage County are not neighbors. Similarly, Geauga and Summit counties are not treated as neighbors.

With regard to the p and q ratios, the analyst often needs to research or look for proper values from theories, literature, historical documents, or past experience

(a) Clustered: 4 BW joins

(c) Dispersed: 8 BW joins

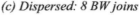

(b) Random:6 BW joins

(d) Number of joins

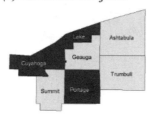

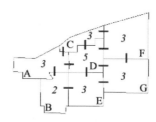

Figure 5.5   Joint structure.

with the subject. For example, if the subject shown by maps in Figure 5.5 is the ratios of infant mortality exceeding a preset threshold at the county level, it would make sense to use the statewide trend to derive proper values for and p and q. Similarly, if the maps in Figure 5.5 show the preferences for either the Democratic or the Republican party at the county level, it may also make sense to research how the two parties fare for the entire state and use that trend to determine appropriate values for and p and q. To illustrate how to calculate E_{BW} and σ_{BW}, let's assume that $p = 0.3$ and $q = 0.7$. Please note that $p + q = 1$ should hold for all cases.

The next step is to calculate the value of $\sum L(L - 1)$. While there may be several ways in which this number can be calculated, we suggest that a table be set up to calculate values of L, then $L - 1$, and then $L(L - 1)$ for each polygon. In this manner, the task of calculating $\sum L(L - 1)$ will be less confusing and easy to manage.

In Figure 5.5d, the polygons are labeled A, B, C, D, E, F, and G. The number of joints for each polygon is its L. As shown in Table 5.5, summing the number of all joints yields $\sum_{i=A}^{G} L_i = 22$. When each L is reduced by 1, the value of $L - 1$ can be easily calculated. The value for $\sum L(L - 1)$ is derived by adding up all $L(L - 1)$ values after multiplying each L and $L - 1$. With the seven polygons in Figure 5.5, $\sum L(L - 1) = 52$, as shown in Table 5.5.

All the necessary values have now been calculated and can be substituted into the following equations:

TABLE 5.5   Joint Count Statistics: Calculating
$\sum L(L - 1)$

	L	$L - 1$	$L(L - 1)$
A	3	2	6
B	2	1	2
C	3	2	6
D	5	4	20
E	3	2	6
F	3	2	6
G	3	2	6
	$\sum L = 22$		$\sum L(L - 1) = 52$

$$E_{BW} = 2 \times 11 \times 0.3 \times 0.7 = 4.62$$

$$\sigma_{BW} = \sqrt{[2 \times 11 + 52] \times 0.3 \times 0.7 - 4 \times [11 + 52] \times 0.3^2 \times 0.7^2}$$

$$= \sqrt{15.54 - 11.11} = \sqrt{4.43} = 2.1.$$

A single number of black-white joints, say 5.04, will not show if there is significant negative spatial autocorrelation in the patterns we have observed. We would
need to compare that number with the expected number of joints in each polygon pattern. This can be done by calculating the difference between the observed
number of black-white joints in each polygon pattern. If we define O_{BW} to be the
observed number of black-white joints in a polygon pattern, we have $O_{BW} = 4$,
6, and 8 for Figure 5.5a, Figure 5.5b, and Figure 5.5c, respectively. With these
numbers, we know that Figure 5.5a is more clustered than a random pattern since
$4 < 4.62$. Figure 5.5b and Figure 5.5c are more dispersed than a random pattern.

The difference between an expected number and an observed number of black-
white joints can help us assess a polygon pattern in terms of its clustering or
dispersion. However, we do not know how far away each pattern is from a random
pattern. Measuring how much each pattern differs from a random pattern requires
that we standardize the calculated difference, $O_{BW} - E_{BW}$.

To standardize the comparison, we will use the Z-score for this procedure.
The Z-score uses the standard deviation of the difference as the denominator to
standardize the difference:

$$Z = \frac{O_{BW} - E_{BW}}{\sigma_{BW}}.$$

From Figure 5.5, we have

$$Z = \frac{4 - 4.62}{2.1} = -0.29 \text{ for Figure 5.5a,}$$

$$Z = \frac{6 - 4.62}{2.1} = 0.65 \text{ for Figure 5.5b,}$$

and

$$Z = \frac{8 - 4.62}{2.1} = 1.61 \text{ for Figure 5.5c.}$$

According to the probability distribution of the Z-score, any Z value that is less than -1.96 or greater than 1.96 is less likely to happen by chance in 5 out of 100 cases ($\alpha = 0.5$). With this information, we can conclude that none of the patterns in Figure 5.5 has a statistically significant negative spatial autocorrelation or a dispersion pattern.

5.5.2   Randomization Sampling

In randomization sampling, the probability of a polygon's being black or white will depend on the total number of black polygons and the total number of white polygons in the polygon pattern being studied. Using Figure 5.5 as an example again, each of the three possible arrangements has exactly three black polygons and four white polygons. The probability of a polygon's being black is $3/7$, and the probability of its being white is $4/7$.

The equations needed to estimate the expected number of joints of all types in nonfree sampling are

$$E_{BB} = \frac{1}{2} W[n_b(n_b - 1)/n(n - 1)]$$

$$E_{WW} = \frac{1}{2} W[n_w(n_w - 1)/n(n - 1)]$$

$$E_{BW} = W[n_b(n - n_b)/n(n - 1)]$$

with standard deviations equal to

$$\sigma_{BB} = \sqrt{\frac{1}{4}\left[\frac{S_1 n(n_b - 1)}{n(n - 1)} + \frac{(S_2 - 2S_1)n_b^{(3)}}{n^{(3)}} + \frac{[W^2 + S_1 - S_2]n_b^{(4)}}{n^{(4)}}\right] - [E_{BB}]^2}$$

$$\sigma_{BW} = \sqrt{\frac{1}{4}\left[\begin{array}{c} \frac{2S_1 n_b n_w}{n(n - 1)} + \frac{(S_2 - 2S_1)n_b n_w(n_b + n_w - 2)}{n^{(3)}} \\ + \frac{4[W^2 + S_1 - S_2]n_b^{(2)} n_w^{(2)}}{n^{(4)}} \end{array}\right] - [E_{BW}]^2}$$

if the stochastic row-standardized matrix is used. σ_{WW} is similar to σ_{BB} defined above, except that n_W is used instead of n_B. Most of these terms were defined in Section 5.4. If the binary connectivity matrix is used instead, the expected value for black-white joints and the corresponding standard deviation are

$$E_{BW} = \frac{2Jn_bn_w}{n(n-1)},$$

and

$$\sigma_{BW} = \sqrt{\begin{array}{l} E_{BW} + \dfrac{\sum L(L-1)n_bn_w}{n(n-1)} \\[2mm] + \dfrac{4[J(J-1) - \sum L(L-1)]n_b^{(2)}n_2^{(2)}}{n^{(4)}} \\[2mm] -E_{BW}^2 \end{array}}.$$

The values of J, $\sum L(L-1)$ are as same as those for the free sampling case. For the polygon patterns in Figure 5.5, the values of E_{BW} and σ_{BW} are

$$E_{BW} = \frac{2 \times 11 \times 3 \times 4}{7 \times 6} = \frac{264}{42} = 6.286$$

and

$$\sigma_{BW} = \sqrt{\begin{array}{l} 6.286 \\[2mm] + \dfrac{52 \times 3 \times 4}{7 \times 6} \\[2mm] + \dfrac{4 \times [(11 \times 10) - 52] \times 3 \times 2 \times 4 \times 3}{7 \times 6 \times 5 \times 4} \\[2mm] -6.286^2 \end{array}}$$

$$= \sqrt{6.286 + 14.857 + 19.886 - 39.514} = \sqrt{1.515} = 1.23.$$

Similarly, we can use the Z-score, $Z = (O_{BW} - E_{BW})/\sigma_{BW}$, to see how each of our sample polygon patterns compares to a random pattern under the condition of three black and four white polygons:

$$Z = \frac{4 - 6.286}{1.23} = -1.85 \quad \text{for Figure 5.5a,}$$

$$Z = \frac{6 - 6.286}{1.23} = -0.23 \quad \text{for Figure 5.5b,}$$

and

$$Z = \frac{8 - 6.286}{1.23} = 1.39 \quad \text{for Figure 5.5c.}$$

Given these Z-scores and the threshold value of ± 1.96, we can conclude that none of the polygon patterns in Figure 5.5a has a significant negative spatial autocorrelation at $\alpha = 0.05$, but Figure 5.5a exhibits a negative spatial autocorrelation lower than a random pattern at $\alpha = 0.1$ (threshold value $= \pm1.645$) level of significance.

ArcView Notes In the project file for this chapter, under the new menu **Spatial Autocorrelation**, there is a menu item for `Joint Count Statistics`. Before one conducts a joint count statistics analysis, a `spatial weights matrix`, either the `binary connectivity matrix` or the `stochastic matrix`, has to be constructed first.

Also, if only selected areal units are included in the construction of the matrix, all spatial autocorrelation procedures should use the same set of selected areal units.

When the procedure for joint count statistics begins, the user will be asked if a weights matrix has been created. If not, the procedure will be terminated.

Otherwise, it will continue to ask for the field used as the identification of areal unit (ID field), and will ask user to select an attribute as the variable for joint count statistics. Be sure that the variable you have chosen is a binary variable.

The ID variable can be alphanumeric. Then the procedure will ask the user to choose either the binary or the stochastic matrix.

After the previously created weights matrix has been selected, the procedure will ask the user to choose a sampling scheme: free sampling or nonfree sampling.

A Report window will be created displaying all related statistics, including observed counts, expected counts, variances, and z-values. The user will be asked if the report should be written to a text file for future use.

5.6 MORAN AND GEARY INDICES

Joint count statistics are useful global measures of spatial autocorrelation for variables with only two outcomes. This situation is quite restrictive, as most real-world cases deal with variables at interval or ratio measurement scales. In these cases, Moran's I and Geary Ratio C can be used.

Moran's I and *Geary's Ratio* have some common characteristics, but their statistical properties are different. Most analysts favor Moran's I mainly because its distribution characteristics are more desirable (Cliff and Ord, 1973, 1981). Still, both statistics are based on a comparison of the values of neighboring areal units. If neighboring areal units over the entire study area have similar values, then the statistics should indicate a strong positive spatial autocorrelation. If neighboring areal units have very dissimilar values, then the statistics should show a strong negative spatial autocorrelation. The two statistics, however, use different approaches to compare neighboring values.

5.6.1   Moran's I

Moran's I can be defined simply as

$$I = \frac{n \sum \sum w_{ij}(x_i - \overline{x})(x_j - \overline{x})}{W \sum (x_i - \overline{x})^2},$$

where x_i is the value of the interval or ratio variable in areal unit i. Other terms have been defined previously. The value of Moran's I ranges from -1 for negative spatial autocorrelation to 1 for positive spatial autocorrelation.

If no spatial autocorrelation exists, the expected value of Moran's I is

$$E_I = -\frac{1}{(n-1)}.$$

When calculating Moran's I, the spatial weights matrices most commonly used are the binary and stochastic matrices. If a binary matrix is used, W in the denominator is basically twice the number of shared boundaries in the entire study region, or $2J$. However, it is possible to use other types of weights matrices. For our purpose, let us assume that a binary matrix is used.

In the numerator of Moran's I, if i and j are neighbors, then w_{ij} will be 1. Therefore, if i and j are not neighbors, the expression will be 0 for that pair of i and j. If they are neighbors, the values of i and j are first compared with the mean of that variable. Their deviations from the mean are then multiplied. The products of the deviations from the mean are then summed for all pairs of areal units as long as they are neighbors. If both neighboring values are above the mean, the product is a large positive number. So is the product if both neighboring values are below the mean (product of two negative numbers).

These situations reflect the presence of positive spatial autocorrelation (i.e., similar values are next to each other). But if the value of one areal unit is above the mean and the value of the neighboring unit is below the mean, the product of the two mean deviations will be negative, indicating the presence of negative spatial autocorrelation. Therefore, over the entire study region, if similar values (can be high-high or low-low) are more likely than dissimilar values between neighbors, Moran's I tends to be positive, and vice versa.

The numerator of Moran's I is based upon the covariance, $(x_i - \overline{x})(x_j - \overline{x})$, which is a cross-product. This covariance structure is also the basis of the Pearson product-moment correlation coefficient, which is defined as

$$r = \frac{\Sigma(x_i - \overline{x})(y_i - \overline{y})}{n\delta_x \delta_y},$$

which measures how closely the distributions of the two variables, x and y, resemble each other. For a given observation, i, if both the x and y values are above their means, the product will be large and positive. Similar result will occur when both the x and y values are below their means. Only if one of the two variables has a value above the mean and the other has a value below the mean will the correlation be negative.

In contrast to the Pearson correlation coefficient, the covariance in Moran's I is the covariance over space for neighboring units and will not be counted unless i and j are neighbors. Also, only one variable is considered instead of two in the Pearson coefficient. The denominator of Moran's I is essentially the sum of the squared deviations scaled by the total weight of the matrix.

Using the seven Ohio counties to illustrate the concepts related to Moran's I, Figure 5.6 shows the variable of the median household income in 1989. Using the binary connectivity matrix and the queen's case to define neighbors, Moran's I is calculated. Tables 5.6a and 5.6b show values derived from the intermediate steps. In Table 5.6a, the median family income (Med Inc), deviations from the mean $(x - \bar{x})$, and the square of the mean deviations $(x - \bar{x})^2$ are reported. The mean of the median household income is \$30,982. Using the mean, the mean deviations were calculated in the third column. The sum of the squared mean deviation, which is $\sum(x_i - \bar{x})^2 = 189,005,048$, is part of the denominator of Moran's I. Another part of the denominator is the sum of the spatial weights, W. Because we use the binary matrix, the sum of all weights is basically $2J$, which is $2 \times 13 = 26$. For the numerator, we have to compute $w_{ij}(x_i - \bar{x})(x_j - \bar{x})$ for each pair of i

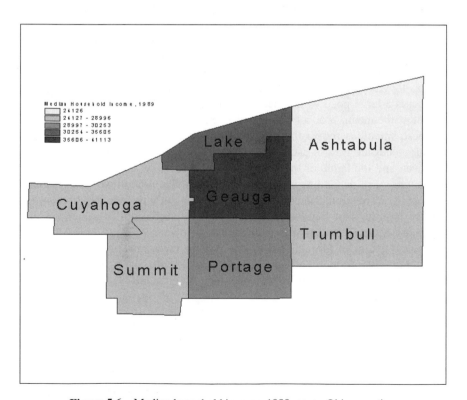

Figure 5.6 Median household income, 1989, seven Ohio counties.

TABLE 5.6a   Mean Deviations and Squared Values of the Mean Deviations

Name	Med Inc	(x − x̄)	(x − x̄)²
Geauga	41,113	10,131	102,637,161
Cuyahoga	28,595	−2,387	5,697,769
Trumbull	28,186	−2,796	7,817,616
Summit	28,996	1,986	3,944,196
Portage	30,253	−729	531,441
Ashtabula	24,126	−6,856	47,004,736
Lake	35,605	4,623	21,372,129
Total	216,874		189,005,048
Mean	30,982		

TABLE 5.6b   Weights Multiplied by the Cross-Products of Mean Deviations

	(x − x̄)	Geauga	Cuyahoga	Trumbull	Summit	Portage	Ashtabula	Lake	Total
(x − x̄)		10,131	−2,387	−2,796	−1,986	−729	−6,856	4,623	
Geauga	**10,131**	0	10,131*(−2,387)	10,131*(−2,796)	10,131*(−1,986)	10,131*(−729)	10,131*(−6,856)	10,131*(4,623)	−102,637,161
Cuyahoga	**−2,387**	−2,387*(10,131)	0	0	−2,387*(−1,986)	−2,387*(−729)	0	−2,387*(4,623)	−28,737,093
Trumbull	**−2,796**	−2,796*(10,131)	0	0	0	−2,796*(−729)	−2,796*(−6,856)	0	−7,118,616
Summit	**−1,986**	−1,986*(10,131)	−1,936*(−2,387)	0	0	−1,986*(−729)	0	0	−13,931,790
Portage	**−729**	−729*(10,131)	−729*(−2,387)	−729*(−2,796)	−729*(−1,986)	0	0	0	−2,159,298
Ashtabula	**−6,856**	−6,856*(10,131)	0	−6,586*(−2,796)	0	0	0	−6,586*(4,623)	−81,984,048
Lake	**4,623**	4,623*(10,131)	4,623*(−2,387)	0	0	0	4,623*(−6,856)	0	4,105,224
								Grand Total =	−232,462,782

and j. The binary weight is reported in Table 5.1. In Table 5.6b, the mean deviations are also listed together with the corresponding counties. As long as the pair of counties has a 1 in the binary matrix, their corresponding mean deviations will be used to form a product reported in Table 5.6b. These cross-products are then summed to give a grand total of $-232,462,782$. Therefore,

$$I = \frac{7^*(-232,462,782)}{26^*189,005,048} = -0.3311.$$

The calculated value for Moran's I seems to indicate a negative spatial autocorrelation. But we have to compare the calculated Moran's I with the expected value. In this example, the expected value is

$$E(I) = \frac{-1}{(7-1)} = -0.16667.$$

Since we do not know if this difference is statistically significant, we have to determine if this difference between the calculated and expected Moran's I values occurred by chance or was very unlikely to happen. To test for the significance of Moran's I, we adopt the same method used to test the significance of joint count statistics.

The difference between the calculated (observed) and expected values of Moran's I is scaled by the standard error of Moran's I in order to derive the z-value. As mentioned before, the expected value of Moran's I is $E_I = -1/(n-1)$, i.e., the value that occurs if there is no spatial autocorrelation. The above definition of expected value for Moran's I applied to any sampling assumptions adopted, but the estimations of variance and standard error vary according to the sampling assumption. Similar to the situation in joint count statistics analysis, there are two sampling assumptions for Moran's I: normality or randomization.

Normality sampling assumes that the attribute values of x_i are independently drawn from a normal distribution and are not limited by the current spatial pattern. Thus, the distribution properties of Moran's I are derived from repeated sampling of a set of values from the normal distribution. Different sets of values and their corresponding means are therefore different. The variance of Moran's I under the normality assumption is

$$\sigma^2(I) = \frac{n^2 S_1 - n S_2 + 3(W)^2}{(W)^2(n^2 - 1)},$$

where all the terms were defined in previous sections.

Under the *randomization* assumption, however, the set of values is fixed. What is not fixed is the location associated with each value. In other words, there are many ways to distribute the set of values in the spatial system. The one we observe is one of many possible spatial patterns given the set of values. The configuration that yields no significant spatial autocorrelation is the one generated by distributing the set of values independently and randomly to areal units. Therefore, the variance is based upon the number of possible permutations of the n data values over the n locations and is defined as

$$\sigma^2(I) = \frac{n[(n^2 - 3n + 3)S_1 - nS_2 + 3W^2] - \left[\frac{1/n \sum(x_i - \bar{x})^4}{[1/n \sum(x_i - \bar{x})^2]^2}\right][S_1 - 2nS_1 + 6W^2]}{(n-1)(n-2)(n-3)(W^2)},$$

where all these terms are defined as before.

Using the seven Ohio counties data and the binary connectivity matrix, $n = 7$, $S_1 = 52$, $S_2 = 416$, and $W = 26$. Therefore, under the normality assumption,

$$\sigma^2(I) = \frac{7^2 * 52 - (7)(416) + 3(26)^2}{(26)^2(7^2 - 1)} = 0.0513.$$

If a randomization assumption is used, $\sigma^2(I) = 0.04945$. The corresponding z-scores are then

$$z_n(I) = -0.3311 - (-0.1667)/\sqrt{0.0513} = -0.7258 \quad \text{for normality}$$

and

$$z_n(I) = -0.3311 - (-0.1667)/\sqrt{0.0495} = -0.7396 \quad \text{for randomization}.$$

Because both z-scores are larger than -1.96 using the traditional criterion (the 0.05 confidence level), both tests indicate that Moran's I is negative. Regardless of which assumption we adopt, we cannot conclude that there is a significant negative spatial autocorrelation in median household income among the seven Ohio counties. The negative Moran's I may be generated by chance; it is not due to a systematic process.

5.6.2 Geary's Ratio

Similar to the Moran's I method of measuring spatial autocorrelation, Geary's Ratio c also adopts a cross-product term (Getis, 1991). Geary's Ratio is formally defined as

$$c = \frac{(n-1) \sum \sum w_{ij}(x_i - x_j)^2}{2W \sum(x_i - \bar{x})^2}.$$

As in Moran's I, Geary's Ratio can accommodate any type of spatial weights matrix, although the most popular types are the binary and stochastic matrices. When this formula is compared with the one for Moran's I, it is apparent that the most significant difference between them is the cross-product term in the numerator. In Moran's I, the cross-product term is based upon the deviations from the mean of the two neighboring values. In Geary's Ratio, instead of comparing the neighboring values with the mean, the two neighboring values are compared with each other directly. To a large degree, we are not concerned about whether x_i is larger than x_j or vice versa, but we are concerned about how dissimilar the two neighboring values are. Therefore, the differences between neighboring values are squared to remove the directional aspect of the differences. The Geary Ratio ranges from 0 to 2, with 0 indicating a perfect positive spatial autocorrelation (i.e., all neighboring values are the same; thus, the cross-product term becomes 0) and 2 indicating a perfect negative spatial autocorrelation. In contrast to Moran's I, the expected value of the Geary Ratio is not affected by sample size n but is always 1.

Using the Ohio counties example, we have all the parameters to calculate the Geary Ratio from the Moran's I calculation except for the sum of the weights times the square of the difference. Table 5.7 shows the derivation of this term. Using the binary connectivity matrix in Table 5.1 again, only if the corresponding cell in Table 5.1 is 1, the corresponding values of the pair of neighbors are compared in Table 5.7. The difference of the value has to be squared too. The sum of the weights times the sum of the squared difference is 2,227,237,238. Then the Geary Ratio is

$$c = \frac{6 * 2,227,237,238}{2 * 26 * 189,005,048} = 1.3697,$$

indicating a slightly negative situation, which is consistent with the result from Moran's I.

As in using Moran's I, we also have to test if the observed Geary Ratio is statistically significant. To derive the z-score, we need to know the expected c and its variance. We know that the expected value of c is 1. For variance estimations under the normality assumption

$$\sigma^2(c) = \frac{(2S_1 + S_2)(n-1) - 4W^2}{2(n+1)W^2},$$

while under the randomization assumption

$$\sigma^2(c) = \frac{\begin{aligned} &(n-1)S_1\left[n^2 - 3n + 3 - (n-1)\left(m_4/m_2^2\right)\right] \\ &-\frac{1}{4}(n-1)S_2\left[n^2 + 3n - 6 - \left(n^2 - n + 2\right)\left(m_4/m_2^2\right)\right] \\ &+W^2\left[n^2 - 3 - (n-1)^2\left(m_4/m_2^2\right)\right] \end{aligned}}{n(n-2)(n-1)W^2}.$$

In our Ohio example, the variances are 0.0385 and 0.0288 for the normality and randomization assumptions, respectively. Therefore, the respective z-scores, based upon the observed minus expected values, are 1.8341 for normality and 2.119 for randomization. Please note that because a 0 in the Geary Ratio means perfect positive spatial autocorrelation and 1 (expected value) means no spatial autocorrelation, a negative z-score means positive spatial autocorrelation and a positive z-score indicates negative spatial autocorrelation. Because the z-score under the randomization assumption and the observed Geary Ratio is

TABLE 5.7 Deriving the Numerator of the Geary Ratio (Weights Multiplied by the Cross-Products of Squared Differences)

	Geauga	Cuyahoga	Trumbull	Summit	Portage	Ashtabula	Lake	Total
	41,113	28,595	28,286	28,996	30,253	24,126	35,605	
Geauga 41,113	X	$(28,595-41,113)^2$	$(28,186-41,113)^2$	$(28,996-41,113)^2$	$(30,253-41,113)^2$	$(24,126-41,113)^2$	$(35,605-41,113)^2$	907,465,175
Cuyahoga 28,595	$(41,113-28,595)^2$	0	0	$(28,996-28,595)^2$	$(30,253-28,595)^2$	0	$(35,605-28,595)^2$	208,750,189
Trumbull 28,186	$(41,113-28,186)^2$	0	0	0	$(30,253-28,186)^2$	$(24,126-28,186)^2$	0	187,863,418
Summit 28,996	$(41,113-28,996)^2$	$(28,595-28,996)^2$	0	0	$(30,253-28,996)^2$	0	0	148,562,539
Portage 30,253	$(41,113-30,253)^2$	$(28,595-30,253)^2$	$(28,186-30,253)^2$	$(28,996-30,253)^2$	0	0	0	126,541,102
Ashtabula 24,126	$(41,113-24,126)^2$	0	$(28,186-24,126)^2$	0	0	0	$(35,605-24,126)^2$	436,809,210
Lake 35,605	$(41,113-35,605)^2$	$(28,595-35,605)^2$	0	0	0	$(24,126-35,605)^2$	0	211,245,605

Grand Total = 2,227,237,238

larger than 1 (negative spatial autocorrealtion), we can conclude that there is a significant negative spatial autocorrelation in median household income among the seven Ohio counties under the randomization assumption. Please note that in this example, the two spatial autocorrelation measures (Moran's I and the Geary Ratio) do not yield consistent conclusions. Even using only one spatial autocorrelation measure (the Geary Ratio) but basing it upon different sampling assumptions gives different results.

ArcView Notes In the project file for this chapter, under the new menu **Spatial Autocorrelation**, there is a menu item for calculating Moran's I and Geary's Ratio. Similar to the procedure for joint count statistics, a spatial weights matrix (binary, stochastic, or distance-based) has to be constructed first. The procedure will ask for the ID field and the variable to be analyzed for spatial autocorrelation. Be sure that the variable you have chosen is an interval or ratio variable. After selecting the weights matrix created previously, the procedure will ask the user to choose a sampling assumption: normality or randomization. If the selected weights matrix is distance-based, the user will be asked to use either a power function or a proportional distance function as the weight. A Report window will be created displaying all related statistics, including the observed and expected statistics, variances, and the *z*-values. This report can be saved as a text file for future use.

5.7 GENERAL G-STATISTIC

Moran's I and Geary's Ratio have well-established statistical properties to described spatial autocorrelation globally. They are, however, not effective in identifying different types of clustering spatial patterns. These patterns are sometimes described as "hot spots" and "cold spots." For instance, if high values are close to each other, Moran's I and Geary's Ratio will indicate relatively high positive spatial autocorrelations. This cluster of high values may be labeled as a hot spot. But the high positive spatial autocorrelation indicated by both Moran's I and Geary's Ratio could also be created by low values close to each other. This type of cluster can be described as a cold spot. Moran's I and Geary's Ratio cannnot distinguish these two types of spatial autocorrelation. The general *G-statistic* (Getis and Ord, 1992) has the advantage over Moran's I and Geary's Ratio of detecting the presence of hot spots or cold spots over the entire study area. These hot spots or cold spots can be thought of as *spatial concentrations*.

Similar to Moran's I and Geary's Ratio, the general G-statistic is also based on the cross-product statistics. The cross-product is often labeled as a measure of spatial association. Formally, the general G-statistic is defined as

$$G(d) = \frac{\sum \sum w_{ij}(d) x_i x_j}{\sum \sum x_i x_j}$$

for $i \neq j$. The G-statistic is defined by a distance, d, within which areal units can be regarded as neighbors of i. The weight $w_{ij}(d)$, is 1 if areal unit j is within d and is 0 otherwise. Thus, the weights matrix is essentially a binary symmetrical matrix, but the neighboring relationship is defined by distance, d. The sum of this weights matrix is

$$W = \sum_i \sum_j w_{ij}(d),$$

where $j \neq i$. Because of this nature of the weight, some of the x_i, x_j pairs will not be included in the numerator if i and j are more than d away from each other. On the other hand, the denominator includes all x_i, x_j pairs, regardless of how far apart i and j are. Apparently, the denominator is always larger than or equal (in the most extreme case using a very large d) to the numerator. Basically, the numerator, which dictates the magnitude of $G(d)$ statistics, will be large if neighboring values are large and small if neighboring values are small. This is a distinctive property of the general G-statistic. A moderate level of $G(d)$ reflects spatial association of high and moderate values, and a low level of $G(d)$ indicates spatial association of low and below-average values.

Before calculating the general G-statistic, one has to define a distance, d, within which areal units will be regarded as neighbors. In the seven Ohio counties, we choose 30 miles for demonstration purposes. Referring to Table 5.8, which shows the centroid distances among these seven counties, 30 miles is large enough for each county to include at least one other county as its neighbor but is not large enough to include all counties for any given county. Based upon the 30-mile criterion used to define neighbors, a binary matrix is derived in Table 5.8. Using median household income as the variable again, the general G-statistic is

$$G(d) = \frac{22,300,327,504}{40,126,136,560} = 0.5557.$$

But the more detailed interpretation of the general G-statistic has to rely on its expected value and the standardized score (z-score).

To derive the z-score and to test for the significance of the general G-statistic, we have to know the expected value of $G(d)$ and its variance. The expected value of $G(d)$ is

$$E(G) = \frac{W}{n(n-1)}.$$

The expected value of $G(d)$ indicates the value of $G(d)$ if there is no significant spatial association or if the level of $G(d)$ is average. In the Ohio case,

TABLE 5.8   Converting a Distance Matrix into a Binary Matrix Using 30 Miles as the Threshold

Distance Matrix Based upon Centroid Distance

ID	Geauga	Cuyahoga	Trumbull	Summit	Portage	Ashtabula	Lake
Geauga	0	25.1508	26.7057	32.7059	25.0389	26.5899	12.6265
Cuyahoga	25.1508	0	47.8151	23.4834	31.6155	50.8064	28.2214
Trumbull	26.7057	47.8151	0	41.8561	24.4759	29.5633	36.7535
Summit	32.7059	23.4834	41.8561	0	17.8031	58.0869	42.7375
Portage	25.0389	31.6155	24.4759	17.8031	0	45.5341	37.4962
Ashtabula	26.5899	50,8064	29.5633	58.0869	45.5341	0	24.7490
Lake	12.6265	28.2214	36.7535	42.7375	37.4962	24.7490	0

Based upon 30 Miles as the Threshold, Distance Matrix Is Converted into Binary Matrix

ID	Geauga	Cuyahoga	Trumbull	Summit	Portage	Ashtabula	Lake
Geauga	0	1	1	0	1	1	1
Cuyahoga	1	0	0	1	0	0	1
Trumbull	1	0	0	0	1	1	0
Summit	0	1	0	0	1	0	0
Portage	1	0	1	1	0	0	0
Ashtabula	1	0	1	0	0	0	1
Lake	1	1	0	0	0	1	0

$$E(G) = \frac{22}{7 \times 6} = 0.5238.$$

Intuitively, because the observed $G(d)$ is slightly higher than the expected $G(d)$, we may say that the observed pattern exhibits some positive spatial association. However, we cannot conclude that this level is significance until we test it. Then we have to derive the z-score of the observed statistic based upon the variance. According to Getis and Ord (1992), the variance of $G(d)$ is

$$\text{Var}(G) = E(G^2) - [E(G)]^2,$$

where

$$E(G^2) = \frac{1}{(m_1^2 - m_2)^2 n^{(4)}} [B_0 m_2^2 + B_1 m_4 + B_2 m_1^2 m_2 + B_3 m_1 m_3 + B_4 m_1^4],$$

where m_j and $n^{(x)}$ were defined in Section 5.4. The other coefficients are as follows:

$$B_0 = (n^2 - 3n + 3)S_1 - nS_2 + 3W^2,$$
$$B_1 = -[(n^2 - n)S_1 - 2nS_2 + 3W^2],$$

$$B_2 = -[2nS_1 - (n+3)S_2 + 6W^2],$$
$$B_3 = 4(n-1)S_1 - 2(n+1)S_2 + 8W^2,$$

and

$$B_4 = S_1 - S_2 + W^2,$$

where S_1 and S_2 were defined in Section 5.4.

The median household income data of the seven Ohio counties give $E(G^2) = 0.2829$. Therefore, the variance of $G(d)$ is

$$\text{Var}(G) = 0.2829 - (0.5238)^2 = 0.0085$$

and the standardized score is

$$Z(G) = \frac{0.5557 - 0.5238}{\sqrt{0.0085}} = 0.3463,$$

which is smaller than 1.96, our standard marker indicating the 0.05 level of significance. In other words, the calculated $G(d)$ has a mild level of spatial association, and the z-score indicates that the counties with high median household income are close to (within 30 miles of) counties with moderate income; this relationship is not statistically significant. That is, the pattern is probably created by chance rather than by some systematic process.

ArcView Notes In the project file for this chapter, under the new menu **Spatial Autocorrelation**, there is a menu item for calculating the general G-statistic. Given the nature of the weight adopted in the G-statistic, only the two distance-based spatial weights matrices are appropriate (centroid distance and distance of the nearest parts). Besides selecting the attribute for the ID field and the variable to be analyzed, the user must enter a distance. Please note that the **View Properties** have to be set correctly for the map unit and the distance unit when constructing the distance matrices. Also, the variable chosen has to be an interval or a ratio variable. A Report window will display all related statistics, including the observed and expected statistics, variances, and the z-values. This report can be exported to a text file for future use.

5.8 LOCAL SPATIAL AUTOCORRELATION STATISTICS

All of the spatial autocorrelation statistics discussed so far share a common characteristic: they are global statistics because they are summary values for the entire study region. It is reasonable to suspect that the magnitude of spatial autocorrelation does not have to be uniform over the region (*spatial homogeneity*), but rather varies according to the location. In other words, it is likely that the magnitude of spatial autocorrelation is high in some subregions but low in other subregions within the study area. It may even be possible to find positive autocorrelation in one part of the region and negative autocorrelation in another part. This phenomenon is called *spatial heterogeneity*.

In order to capture the spatial heterogeneity of spatial autocorrelation, we have to rely on another set of measures. All these measures are based upon their global counterparts discussed above (Moran's I, the Geary Ratio, and the general G-statistic) but are modified to detect spatial autocorrelation at a local scale.

5.8.1   Local Indicators of Spatial Association (LISA)

Local Indicators of Spatial Association refer to the local version of Moran's I and Geary Ratio (Anselin, 1995). In order to indicate the level of spatial autocorrelation at the local scale, a value of spatial autocorrelation has to be derived for each areal unit. The local Moran statistic for areal unit i is defined as

$$I_i = z_i \sum_i w_{ij} z_j,$$

where z_i and z_j are in deviations from the mean or

$$z_i = (x_i - \overline{x})/\delta$$

and δ is the standard deviation of x_i. Similar to the interpretation of Moran's I, a high value of *local Moran* means a clustering of similar values (can be high or low), while a low value of local Moran indicates a clustering of dissimilar values. In general, w_{ij} can be the row-standardized matrix, but other spatial weights matrices are also appropriate. If the weight is in row-standardized form, the local Moran for areal unit i is basically the mean deviation of i multiplied by the sum of the products of the mean deviations for all j values and the spatial weights defining the spatial relationship between i and j.

Table 5.9 is similar to Table 5.6. From Figure 5.1 and Table 5.1, we can see that Geauga County has six neighboring counties. If the row-standardized stochastic weights are adopted, the weight for each of Geauga County's neighbors is $1/6$. The row-standardized weights are reported in Table 5.9a. The number of deviations from the mean value of median household income for all seven counties is reported in Table 5.9b. Among the seven Ohio counties, let us focus on Geauga County. Because Geauga County has six neighbors, the weight for each neighbor is $1/6$. Applying this weight to the mean deviations for all the neighbors of Geauga County defined by the 30-mile criterion, $w_{ij} \cdot z_{ij}$ for each neighbor is reported. The same step is applied to other counties. In Table 5.9b, the products of the weight and mean deviation for all neighboring counties are summed and then multiplied by the mean deviation of the county of concern. That gives us the local

TABLE 5.9 Computing Local Indicators of Spatial Association Using the Seven Ohio Counties

(a) Row-Standardized Weight Matrix for the Seven Ohio Counties

ID	Geauga	Cuyahoga	Trumbull	Summit	Portage	Ashtabula	Lake
Geauga	0	1/6	1/6	1/6	1/6	1/6	1/6
Cuyahoga	1/4	0	0	1/4	1/4	0	1/4
Trumbull	1/3	0	0	0	1/3	1/3	0
Summit	1/3	1/3	0	0	1/4	0	0
Portage	1/4	1/4	1/4	1/4	0	0	0
Ashtabula	1/3	0	1/3	0	0	0	1/3
Lake	1/3	1/3	0	0	0	1/3	0

(b) Stochastic Weights Multiplied by Deviations from the Mean

	Geauga	Cuyahoga	Trumbull	Summit	Portage	Ashtabula	Lake	Total (wz)	Li
(X-mean)/std	1.949687	−459372	−0.538083	−0.382201	−0.140294	−1.319421	0.889685		
Geauga	0	1/6 * −0.459372	1/6 * −0.538083	1/6 * −0.382201	1/6 * −0.140294	1/6 * −1.319421	1/6 * 0.889685	−102,637.161	−0.633546
Cuyahoga	1/4 * 1.949687	0	0	1/4 * −0.382201	1/4 * −0.140294	1/4 * −1.319421	1/4 * 0.889685	−28,737.093	−0.266077
Trumbull	1/3 * 1.949687	0	0	0	1/3 * −0.140294	1/3 * −1.319421	0	−7,118.616	−0.087882
Summit	1/3 * 1.949687	1/3 * −0.459372	0	0	1/3 * −0.140294	0	0	−13,931.790	−0.171993
Portage	1/4 * 1.949687	1/4 * −0.459372	1/4 * −0.538083	1/4 * −0.382201	0	0	0	−2,159.298	−0.019993
Ashtabula	1/3 * 1.949687	0	1/3 * −0.538083	0	0	0	1/3 * 0.889685	−81,984.048	−1.012123
Lake	1/3 * 1.949687	1/3 * −0.459372	0	0	0	1/3 * −1.319421	0	4,105.224	0.050681

(c) LISA and Related Statistics

Counties	Li	EXP(Li)	Var(Li)	Z(Li)	Ci
Geauga	−0.63367	−0.1667	0.07609	−1.69293	5.60261
Cuyahoga	−0.26608	−0.16667	0.13911	−0.26652	1.93282
Trumbull	−0.08787	−0.16665	0.20131	0.17558	2.31901
Summit	−0.17198	−0.16665	0.20296	−0.01182	1.83387
Portage	−0.01999	−0.16667	0.13911	0.39325	1.17165
Ashtabula	−1.01202	−0.16665	0.19966	−1.89193	5.39202
Lake	0.05068	−0.16665	0.20121	0.48437	2.60764

Moran for each of the seven counties. The major difference among the counties is their weights, depending on the number of neighbors each of them has.

As with other statistics, just deriving the local Moran values for each county is not very meaningful. High or low local Moran values may occur just by chance. These values have to be compared with their expected values and interpreted with their standardized scores. According to Anselin (1995), the expected value under the randomization hypothesis is

$$E[I_i] = -w_{i.}/(n-1),$$

and

$$\text{Var}[I_i] = w_{i.}^{(2)} \frac{(n - m_4/m_2^2)}{(n-1)} + 2w_{i(kh)} \frac{(2m_4/m_2^2 - n)}{(n-1)(n-2)} - \frac{w_{i.}^2}{(n-1)^2},$$

where

$$w_{i.}^2 = \left(\sum_j w_{ij} \right)^2$$

and

$$w_{i.}^{(2)} = \sum_j w_{ij}^2; \quad i \neq j.$$

The term

$$2w_{i(kh)} = \sum_{k \neq i} \sum_{h \neq i} w_{ik} w_{ih}.$$

Table 5.9c reports the expected values, variances, and z-scores for all seven counties. Note that because each county has its own local Moran, each local Moran has its associated expected value and variance. This is an advantage of the local Moran. A value is derived for each areal unit, and therefore the results can be mapped. Figure 5.7 consists of two maps: the local Moran values and the z-scores for all seven counties. The local Moran reflects how neighboring values are associated with each other. From Figure 5.6, we can see that the median household income of Geauga County is highest and that one of its neighboring counties, Ashtabula County, has the lowest income. Therefore, the local Moran for Geauga County is rather low and in fact negative. But Ashtabula County has the lowest local Moran because it is very different from its surrounding counties, including Geauga County. On the other hand, Cuyahoga and Summit counties are similar in their relatively low income levels, and therefore their local Moran values are moderately high (similar low values are next to each other). Lake and Portage Counties, and to some extent Trumbull County, surrounding Geauga County, have relatively high or moderate income levels. Therefore, these three counties have

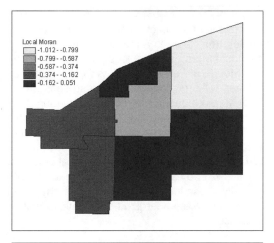

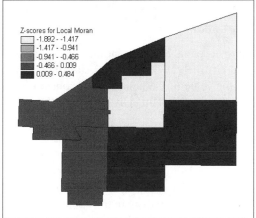

Figure 5.7 Local Moran of median household income, seven Ohio counties.

very high local Moran values, indicating that neighboring units have rather similar values. Still, none of the standardized local Moran scores exceeds the ± 1.96 range. Thus, the pattern we observe may be the outcome of a random process.

A local version of the Geary Ratio is also available. Formally, it is defined as

$$c_i \sum_j w_{ij}(z_i - z_j)^2.$$

Unfortunately, the distribution properties of the local Geary are not as desirable as local Moran. Still, mathematically, we can interpret the local Geary the same way as the global Geary Ratio. Clustering of similar values will create a relatively low local Geary, while clustering of dissimilar values will yield a relatively high local Geary. Similar to the Moran, a local Geary value is computed for each areal unit; therefore, the results can be mapped.

ArcView Notes In the project file for this chapter, under the new menu **Spatial Autocorrelation**, there is a menu item for calculating the LISA. Both the local Moran and the local Geary will be calculated. For the local Moran, all related statistics for significance testing and the z-scores will be calculated. Because each areal unit has its own local Moran measure, to facilitate mapping of the statistics, the statistics together with the z-scores will be written to the attribute table for future mapping. After the LISA procedure is completed, the user can map these statistics using the standard ArcView mapping process. But for the local Geary, only the local statistics will be calculated and written to the attribute table. Theoretically, these local statistics support any type of weights matrix, but the row-standardized stochastic matrix is the most logical choice.

5.8.2 Local G-Statistics

Another local measure of spatial autocorrelation is the local version of the general G-statistic (Getis and Ord, 1992). The local G-statistic is derived for each areal unit to indicate how the value of the areal unit of concern is associated with the values of surrounding areal units defined by a distance threshold, d. Formally, the local G-statistic is defined as

$$G_i(d) = \frac{\sum_j w_{ij}(d)x_j}{\sum_j x_j}; \quad j \neq i.$$

All other terms were defined previously in the discussion of the general G-statistic. It is best to interpret the statistic in the context of the standardized score. To obtain the standardized score, we need to know the expected value and the variance of the statistic. The expected value is defined as

$$E(G_i) = W_i/(n-1),$$

where

$$W_i = \sum_j w_{ij}(d).$$

The definition of the variance is similar to the definition of the general G-statistic. It is defined as

$$\text{Var}(G_i) = E(G_i^2) - [E(G_i)]^2$$

and

$$E(G_i^2) = \frac{1}{(\sum_j x_j)^2} \left[\frac{W_i(n-1-W_i)\sum_j x_j^2}{(n-1)(n-2)} \right] + \frac{W_i(W_i-1)}{(n-1)(n-2)},$$

where $j \neq i$. Given the standardized score of $G_i(d)$ using the above expected value and variance, a high score appears when the spatial clustering is formed by similar but high values. If the spatial clustering is formed by low values, the z-score will tend to be highly negative. A z-score around 0 indicates no apparent spatial association pattern. A related statistic is labeled $G_i^*(d)$. This statistic is almost identical to $G_i(d)$, except that it includes cases where $j = i$. Because these two statistics are so similar, we will focus on $G_i(d)$. Readers who are interested in the other statistic can refer to Getis and Ord (1992).

Given the interpretation of the local G-statistic, we should expect that, using median household income as the variable, Geauga County will have a high local G-statistic because its two neighbors, Lake and Portage counties, have relatively high values. As shown in the upper map in Figure 5.8 and Table 5.10, the local G-statistic for Geauga County is the highest. Summit County has the lowest local G-statistic because both of its neighbors (defined by the 30-mile distance selected

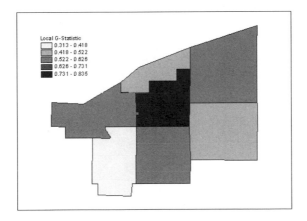

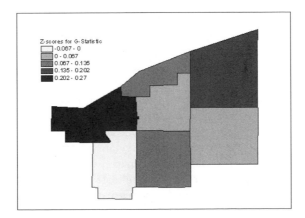

Figure 5.8 Local G-statistics and their Z-scores, seven Ohio counties.

TABLE 5.10 Local G-Statistics and Related Statistics

County	Gi	E(Gi)	VAR(Gi)	Z(Gi)
Geauga	0.83503	0.83333	0.00609	0.02169
Cuyahoga	0.56148	0.50000	0.05198	0.26965
Trumbull	0.50608	0.50000	0.05487	0.02597
Summit	0.31322	0.33333	0.08899	−0.06741
Portage	0.52671	0.50000	0.05326	0.11574
Ashtabula	0.54425	0.50000	0.05229	0.19352
Lake	0.51765	0.50000	0.05528	0.08507

a priori) have relatively low levels of income. It is interesting that for Lake County, the three neighbors within 30 miles (Cuyahoga, Geauga, and Ashtabula counties) have very different income levels. As a result, the local G-statistic for Lake County is moderately negative, indicating moderate negative spatial autocorrelation.

ArcView Notes Under the new menu **Spatial Autocorrelation**, there is a menu item for calculating the local G-statistic. The user has to provide a distance (based on the distance unit adopted to create the distance matrix) to define neighboring units. The local G-statistic and its related statistics are calculated and written into the attribute table for future mapping.

5.9 MORAN SCATTERPLOT

The development of local spatial autocorrelation statistics acknowledges the fact that spatial processes, including spatial autocorrelation, can be heterogeneous. Apparently, we can adopt one or more of the previously described local measures of spatial autocorrelation to evaluate the entire study area if there is any local instability in spatial autocorrelation. From the statistical visualization and spatial exploratory analysis perspectives, it will be informative and useful if the analyst can identify areas with unusual levels of spatial autocorrelation. Those areas can be regarded as the outliers. A very effective visual diagnostic tool is the Moran scatterplot based upon a regression framework and Moran's I statistic (Anselin, 1995). Assuming that x is a vector of x_i, the deviation from the mean $(x_i - \overline{x})$ and W is the row-standardized spatial weights matrix, we may form a regression of Wx on x, while the slope of this regression indicates how the neighboring values are related to each other. In other words, the regression is

$$x = a + IWx,$$

where a is a vector of the constant intercept term and I is the regression coefficient representing the slope. The slope is therefore also the Moran's I global statistic.

Moran's I reflects the level of spatial autocorrelation, and the statistic is a global summary statistic. Different observations within the study region, however, may show different levels of spatial autocorrelation with neighbors. By plotting Wx on x superimposing the regression line, the scatterplot can potentially indicate outliers in term of the magnitude of spatial autocorrelation. If all observations have a similar level of spatial autocorrelation, the scatterplot will show observations lying close to the regression line. If certain observations show unusually high or low levels of spatial autocorrelation locally in reference to their neighbors, those observations will be plotted far above or below the regression line. This regression line reflects the general trend of spatial autocorrelation in the entire region, and the slope parameter is equivalent to Moran's I. In other words, those observations deviating from the general trend of spatial autocorrelation have spatial autocorrelations that are very different from the overall level. Thus, the Moran scatterplot is useful in identifying unusual observations in regard to relationships with neighbors.

In the Ohio case, the median household income values among the seven counties have moderately negative spatial autocorrelation. High-value counties are close to low-value counties. This finding is the result of using a few counties for demonstration. The scatterplot in Figure 5.9 shows that the income (x) is in-

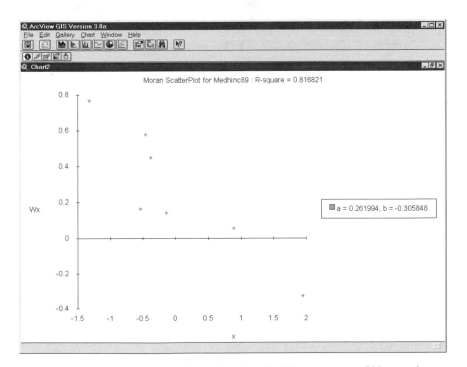

Figure 5.9 Moran scatterplot for median household income, seven Ohio counties.

versely related to the product of the spatial weights and income (Wx). The slope is -0.3058, which is moderately negative and not too diffferent from the Moran's I using the binary weights matrix (-0.3311). Please note that the value of the slope parameter is identical to the Moran's I of median household income when a row-standardized weights matrix is used. The R-square turns out to be quite high, mainly because of a few observations. The extremes are Geauga County, as shown by the point in the lower right corner, and Ashtabula County, as indicated by the point in the upper left corner. The scatterplot basically shows that the high income level of Geauga County (x) is negatively associated with the low income levels of surrounding counties. By contrast, the low level income of Ashtabula County contrasts to the high income levels of its surrounding counties. Because only seven observations are involved, the scatterplot does not reveal any obvious outliers. Figure 5.10 provides another example using 49 states in the continental United States and their median house values. The R-square is lower than in the Ohio case (0.4732), but the slope of Moran's I is 0.5526, indicating moderately positive spatial autocorrelation. In this case, the median house value of a state is positively and moderately associated with the house value levels of the surrounding states. An outlier, however, is California, the state close to the x-axis at the extreme right. The deviation of California from the upward-sloping pattern in-

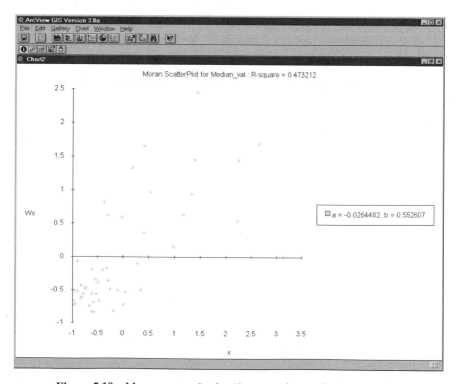

Figure 5.10   Moran scatterplot for 49 states using median house value.

dicates extremely high house value, in contrast to the values of the surrounding states.

ArcView Notes Under the new menu **Spatial Autocorrelation**, there is a menu item for creating a Moran scatterplot. The scatterplot requires a row-standardized stochastic weights matrix. The procedure will ask for the matrix and the variable to be analyzed. The regression coefficients (the intercept and the slope) will be displayed in a window. The deviations from the mean (x) and the weight times the mean (Wx) will be written to the attribute table in Arc-View for plotting a chart. The procedure will continue to ask the user if a scatterplot should be created. If the answer is yes, a Moran scatterplot will be created in the **Chart** window together with the regression parameters using values in the two new fields written to the attribute table. Please note that if the identification tool in ArcView is chosen, the user can click on the observation (a point) in the plot to find out which observation the point represents.

5.10 APPLICATION EXAMPLES

The methods discussed in this chapter are used to calculate spatial autocorrelation in various forms. The joint count statistic procedure is used with nominal or binary data. Moran's I, Geary's Ratio, and G-statistics are for overall evaluation of spatial patterns, while local indicators and local G-statistics measure local variation in spatial autocorrelation. Finally, the Moran scatterplot is a tool to visually identify unusual levels of spatial autocorrelation among areal units.

In working with these methods for this example, we will use the U.S. county database. Data themes have been extracted for the states of Ohio, Indiana, Illinois, and Kentucky. We will examine the spatial patterns of one variable, the median housing value, in these four states.

There is a median housing value for each county in each of the four states. First, we calculate the average of all median housing values in each state as a cutoff point to create a binary category. Those counties with median housing values less than the average in their state will be assigned a value of 1. Those counties with median housing value greater than the state's average will be assigned a value of 0.

Below we describe the steps for calculating each index of spatial autocorrelation. The general structure for processing each data theme is as follows:

1. Copy the shapefiles from the companion website to this book to your working directory. This is needed so that we can make changes in these files.

2. Calculate descriptive statistics of the chosen variable, `Median_val`. We will use the average of all median housing values as the cutoff point, setting those counties with higher median housing values equal to 0 and those counties with lower median housing values equal to 1.

3. Edit the attribute table to add a new field for the new binary values of 1s and 0s as defined above.

4. Calculate weight matrices and save them in the working directory.

5. Calculate the joint count statistic using the newly added binary field.

6. Calculate Moran's I and Geary's C.

7. Calculate the global G-statistic.

8. Calculate the local indicators for each county.

9. Calculate the local G-statistics for each county.

10. Construct a Moran scatterplot.

11. Repeat Steps 1 to 10 for each state.

When the procedure is completed, we will be able to examine all the results and compare them for the four states. At this point, it is necessary to point out that the more counties (polygons) we have in a data theme, the more time will be needed to complete the calculation. The time required depends on the computer's speed as well as the distance measures selected.

5.10.1 Preparation

To use the various spatial autocorrelation indices, we need to prepare the data theme and its attribute table. To do so, follow these steps:

1. Start the ArcView and open the project file, `Ch5.apr`, downloaded from the website.

2. Use **View/Add Theme** from the View document's menu to add the `oh.shp` downloaded from the website.

3. Because we are going to make changes on the themes, if you want to keep the original copy of the theme, you must create a duplicate. Use the **Theme/Convert to Shape File** menu item from the View document. Save **oh.shp** in your working directory. When asked if the theme is to be added to the View document, click the **Yes** button. The original **oh.shp** can be deleted from the Table of Contents in the View document by clicking it (to make it active) and then using **Edit/Delete Theme** to remove it from the View document.

Next, we will add a binary attribute field, using the variable being analyzed. First, we will calculate descriptive statistics for this variable and then use its mean to create a new attribute field, `hprice`.

1. Use **Theme/Table** from the View document menu to open the associated attribute table.

2. In the Table document, **Attributes of Oh.shp**, scroll to the right until the field of median housing value (**Median_val**) appears.

3. Click the title button of **Median_val** to highlight this field. Use the **Field/Statistics...** menu item to calculate the descriptive statistics of this field. We see that the average of all 88 counties' median housing value is $54,884. Figure 5.11 shows the descriptive statistics derived from the median housing values.

4. In the Table document, **Attributes of Oh.shp** (make the table active), use **Table/Start Editing** from the menu to begin editing. Figure 5.12 shows the screen when setting the Table for editing.

5. Use **Edit/Add Field** from the Table document's menu to invoke the **Field Definition** window. In it, change the name of the field to be hprice, type to be number, width to be 2, and decimal places to be 0. Notice that a new field is created and highlighted.

6. Use the **Table/Query** menu item to open the query window named **Attributes of Oh.shp**. From the list of fields, scroll down and double click

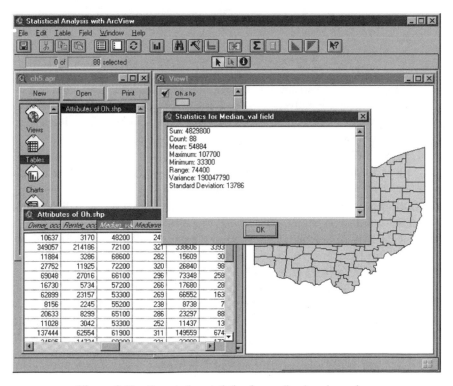

Figure 5.11   Descriptive statistics for median housing values.

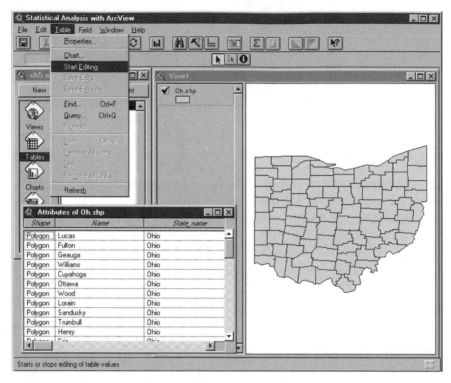

Figure 5.12 Edit table document.

[Median_val] to add it to the query. Next, click the "<" button to add it to the query. Finally, type 54884 to complete the query. Now click the **New Set** button to select all counties whose median housing value is less than $54,884.

7. Use the **Field/Calculate...** menu item to open the **Field Calculator** window. In the [hprice]= box, type 1 to set the value of hprice for all selected records to be 1. Click **OK** to proceed.

8. To assign another value to the unselected records, use the **Edit/Switch Selection** menu item to select the counties that have not yet been assigned values for hprice. When these are selected, use the **Field Calculator** to assign them a value of 0 by repeating Step 7 but typing 0 in the [hprice]= box.

9. Unselect all records so that the subsequent procedures are applied to all records. To do so, use the **Edit/Select None** menu item.

10. Use **Table/Stop Editing** to finish the procedure of adding the binary field. When asked whether you wish to save edits, click the Yes button to save the changes.

With the data theme prepared, the next task is to calculate the weight matrices.

5.10.2 Weight Matrices

Adding the binary field allows the calculation of joint count statistics. In calculating weight matrices, the binary field is directly related to the weight matrix.

The procedure for calculating various weight matrices is as follows:

1. From the View document's menu, use the **Spatial Autocorrelation/ Creating Weight Matrices** menu item to start the process of creating a binary weight matrix. Figure 5.13 shows the added user interface in the project file, Ch5.apr.

2. In the **Info** window, click **OK** to confirm that no feature was selected at this point.

3. In the **Selection** window, click the **Yes** button to agree to select all records.

4. In the **ID Selection** window, choose Cnty_fips as the ID field and then click **OK** to proceed.

5. In the **Distance Definition** window, choose Binary Connectivity as the distance measure and then click **OK** to proceed.

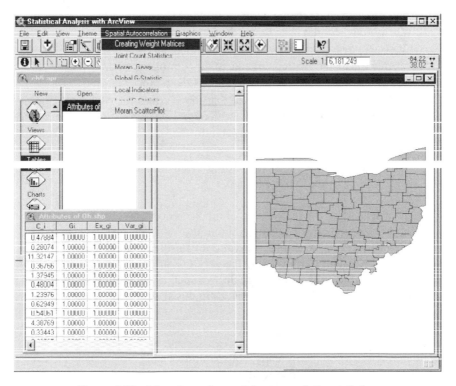

Figure 5.13 Menu items for spatial autocorrelation statistics.

6. When prompted, navigate to the working directory. Give `ohbinary.dbf` as the name of the file. Click **OK** to proceed. In the subsequent **Info** window, click **OK** to acknowledge the creation of the distance matrix.

7. Repeat the steps to create other matrix files by selecting different distance measures in the **Distance Definition** window. Save the output files using the names `ohstoch.dbf` for **Stochastic Weight**, `ohnparts.dbf` for **Distance between Nearest Parts**, and `ohcdist.dbf` for **Centroid Distance**. Store all of these weight matrices in the working directory.

5.10.3   Joint Count Statistics

Joint count statistics use binary variables and either a binary connectivity matrix or a stochastic matrix. To calculate joint count statistic, follow these steps:

1. Use the **Spatial Autocorrelation/Joint Count Statistics** menu item to start the process. In the **Check for input** window, click **Yes** to indicate that the matrix has been created.

2. In the **Get Input** window, choose `Cnty_fips` as the ID field. Click **OK** to proceed.

3. In the next **Get Input** window, scroll down the list and select `hprice` as the variable. Click **OK** to continue.

4. When prompted, navigate to the directory and select `ohbinary.dbf` as the weight matrix file if the binary matrix is used.

5. When the calculation is completed, in the **Sampling** window, choose `Normality` as the sampling scheme. Click **OK** to proceed.

6. From the **Report** window, we see that the z-values for AA and AB joints are all statistically significant (> 1.96 or < 1.96) while the z value for BB joints is not statistically significant. Figure 5.14 shows the results of running joint count statistics.

7. In the **Write txt file** window, click the **Yes** button to save the results. Give `ohjcs.txt` as the text file name.

As the results show, the distribution of median housing values is far from random. In fact, we can conclude that median house value has a significant positive spatial autocorrelation, because the z-value of positive autocorrelation (AA) is larger than 1.96, which is affirmed by the very low negative z-value for AB. But this is measured only at the nominal scale. We shall use other indices to examine the spatial pattern of median housing values in this state.

5.10.4   Moran-Geary Indices

While they are structured differently, both Moran's I and Geary's Ratio are useful measurements of spatial autocorrelation. To calculate their index values, follow these steps:

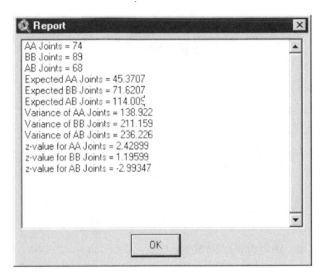

Figure 5.14 Results of joint count statistics analysis.

1. From the View document's menu, use the **Spatial Autocorrelation/Moran-Geary** menu item to start.

2. In the **Check for Input** window, click the **Yes** button to confirm the existence of a spatial weight matrix.

3. In the **Get Input** window, select Cnty_fips as the ID field.

4. In the next **Get Input** window, scroll down to select Median_val as the variable to be analyzed.

5. In the **Choose Spatial Weight Matrix File** window, navigate to the directory and select ohstoch.dbf. At this point, the selection of the spatial weighting matrix file is not limited to the use of ohstoch.dbf or ohbinary.dbf. Users may choose to use ohcdist.dbf instead. However, the selection of the matrix file should also be reflected in the selection of the next window.

6. In the **Distance Definition** window, select Stochastic Weight as the distance measure used to create the matrix file. Again, if you are using a weight matrix file other than ohstoch.dbf, the selection of the distance measure should reflect the matrix file used.

7. From the **Report** window, we see that Moran's I is 0.449089, with a z-value of 6.94788 under normality sampling, and Geary's C is 0.56978, with a z-value of -6.42476 under Normality sampling. Figure 5.15 shows the results from Moran-Geary menu item.

8. Save the result as ohmg.txt in the directory.

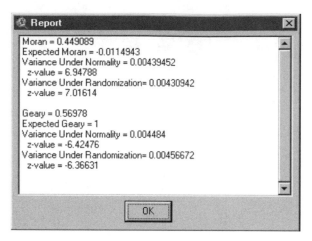

Figure 5.15   Output from the Moran-Geary spatial autocorrelation menu item.

The values of the two indices indicate strong spatial autocorrelation. They confirm the conclusion from joint count statistics analysis that the pattern is not random.

5.10.5   Global G-Statistics

Other measures that can be used to measure the global pattern of spatial autocorrelation are the G-statistics. In applying them to the Ohio county data theme, the procedures are as follows:

1. From the View document's menu, use **Spatial Autocorrelation/Global G-Statistics** to start.

2. In the **Distance Entry** window, enter 50 (miles) as the search distance for defining neighbors.

3. In the **Get Input** window, select Cnty_fips as the ID field.

4. In the next **Get Input** window, select Median_val as the variable to be analyzed.

5. When prompted, choose ohcdist.dbf as the centroid distance matrix file from the directory.

6. Upon completion of the calculation, the statistics are reported in the **Result** window. We can see that the calculated value is 0.144867, with a z-value of 1.7808, not statistically significant at the 95% level. Figure 5.16 shows the output from running G-statistics.

7. When prompted, save the result in the same directory under the name ohgg.txt.

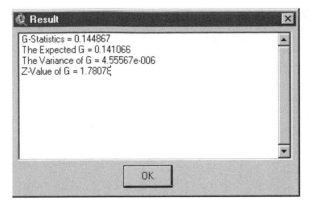

Figure 5.16   Output from the global G-statistics autocorrelation menu item.

5.10.6   Local Indicators

To examine how a spatial pattern changes within itself, we will apply the local indicators. To do so, follow these steps:

1. From the View document's menu, use **Spatial Autocorrelation/Local Indicator** to begin.
2. Click the **Yes** button in the **Check for input window** to confirm the creation of a weight matrix file.
3. Choose Median_val as the variable to be analyzed.
4. Choose ohstoch.dbf as the stochastic weight matrix file.
5. When the calculation is done, click the Table document, **Attributes of Oh.shp**, to make it active. We see that four attributes have been created: I_i, Exp I_i, Val_I_i, Z I_i, and C_i. These fields give the values of local indicators, their expected values, variances, z-values, and so on. If desired, the user can double click the **Oh.shp** theme in the View document to invoke the Legend Editor. Specify **Graduated Color** to be the Legend Type and I_i to be the Classification Field. Click the **Apply** button to display the spatial pattern of the calculated local indicators. Figure 5.17 gives an example for mapping the local indicators.

It may be necessary to adjust the classification in the Legend Editor or the color scheme to see how these local indicators display the changes.

5.10.7   Local G-Statistics

Other local indices are the local G-statistics. To calculate them for the counties, follow these steps:

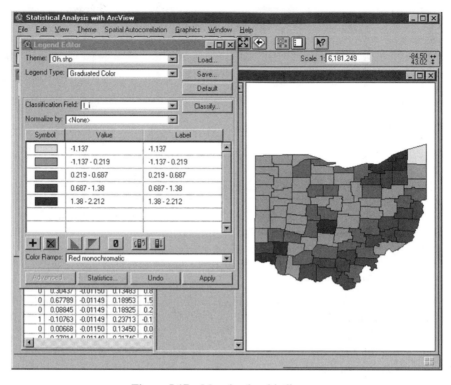

Figure 5.17 Mapping local indicators.

1. From the View document's menu, use **Spatial Autocorrelation/Local G-Statistics** to continue.
2. In the **Distance Entry** window, type 50 as the distance in miles.
3. In the **Get Input** window, choose Cnty_fips as the ID field.
4. In the next **Get Input** window, choose Median_val as the variable to be analyzed.
5. When prompted, choose ohcdist.dbf as the centroid distance matrix file.
6. When the calculation is complete, click the Table document to make it active. Again, scroll to the right to see the newly created fields: Gi, Ex_Gi, Var_Gi, and Z_Gi, similar to the output from local indicators. These fields give values of local G-statistics, their expected values, their variances, and their z-values.

If desired, the Legend Editor can be used to map the distribution of local G-statistics.

5.10.8 Moran Scatterplot

One way to see the index values graphically is to construct a scatterplot to display them. We can use the Moran scatterplot menu to do so:

1. From the View document's menu, use the **Spatial Autocorrelation/Moran ScatterPlot** menu item to begin.
2. Confirm the creation of a stochastic weight matrix by clicking the **Yes** button in the next window. A stochastic matrix is needed for Moran scatterplot.
3. Choose `Cnty_fips` as the ID field.
4. Choose `ohstoch.dbf` as the weight matrix file.
5. In the **Bivariate Regression Results** window, we see that the **R-squared** value is 0.4145, with the intercept of the regression line being 0.0258799 and the slope being 0.449148. Click **OK** to proceed. Note that the slope is the same as the Moran's I reported earlier.
6. In the **Bivariate Regression** window, click the **Yes** button to create a Moran scatterplot.
7. When the plot is created, expand the **Chart1** window horizontally to give the scatterplot more space for the x-axis. Figure 5.18 shows an example of the results.

In Moran scatterplots, the x-axis is the original value (deviation from the mean) and Wx is the predicted x value based on neighboring values. The scatterplots show regression lines between the two axes. The slope of the regression line is identical to that of Moran's I using the stochastic weight matrix. If the slope runs from lower left to upper right, the spatial autocorrelation is positive. The spatial autocorrelation is negative if the slope runs from upper left to lower right.

Because the magnitude of spatial autocorrelation over a region is not uniform, to determine the spatial heterogeneity of spatial autocorrelation, the scatterplot can show how closely each areal unit aligns with the regression line. The model

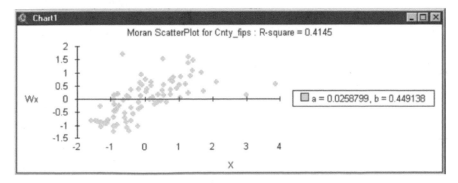

Figure 5.18   Moran scatterplot.

is summarized by the R-square value that indicates how well the regression represents the distribution. If all areal units have a similar magnitude of spatial autocorrelation, they should lie very close to a straight line. However, outliers can be easily identified.

5.10.9 Comparing States

Once the procedures for calculating various spatial autocorrelation statistics are completed, they can be repeated for Indiana, Illinois, and Kentucky. The results of these calculations are shown in Table 5.11.

From the average of median housing values in the counties of each state, we can see that Ohio has the highest housing values, followed by Indiana, Illinois, and Kentucky. The Z-values for joint count statistics among the four states show various degrees of departure from a random pattern. For example, Ohio has more AA joints but fewer AB joints than a random pattern. For Indiana, only AB joints are different from a random pattern with statistical significance. Illinois and Kentucky have patterns similar to that of Ohio.

With Moran's I and Geary's Ratio, all four states show spatial patterns with statistical significance. Illinois has the strongest spatial pattern. With the largest Moran's I and the smallest Geary's Ratio, Illinois has the most clustered pattern, in which adjacent polygons show similar median housing values. This finding is further verified by the global G-statistics. Furthermore, Illinois has the highest R-square value in the Moran scatterplot, with the steepest slope for the regression line.

TABLE 5.11   Spatial Autocorrelation Statistics for Counties in Ohio, Indiana, Illinois, and Kentucky Using Median Housing Values

	Ohio	Indiana	Illinois	Kentucky
Average of median housing values	$54,884	$48,715	$47,098	$41,393
AA joins	74	61	53	73
Z(AA) value	2.4289	1.1005	2.4540	1.7234
BB joins	89	91	141	130
Z(BB) value	1.1959	1.1650	1.2084	1.2473
AB joins	68	88	75	106
Z(AB) value	−2.9935	1.9537	−2.5123	−2.3975
Moran Index	0.4491	0.3144	0.7471	0.5000
Z(Moran)	6.9479	4.9828	12.2387	8.7378
Geary's Ratio	0.5698	0.6667	0.1986	0.4739
Z (Geary)	1.7808	1.9137	2.8359	4.1025
Moran scatterplot				
R^2	0.4145	0.3122	0.8280	0.5341
a (intercept)	0.0259	−0.0073	−0.0103	0.0009
b (slope)	0.4491	0.3144	0.7471	0.5000

5.11 SUMMARY

In real-world GIS applications, spatial data are often treated as data without a spatial dimension. Classical statistical techniques, which often assume that the observations are independent of each other, are used indiscriminately on spatial data as if they are ordinary data. The unique characteristics of spatial data are ignored, and the analytical results may be biased. In this chapter, we have argued that a unique characteristic of spatial data is (spatial) dependency, and evaluating spatial dependency, or autocorrelation, is an important step in analyzing spatial data. If the assumption of independence is violated, classical statistical techniques for drawing inferences will be inappropriate.

Different types of spatial autocorrelation and measures have been discussed in this chapter. Global measures are summary measures for the entire region, while local measures depict the situation for each areal unit. Some measures are effective in identifying spatial trends, while others are efficient in distinguishing hot spots from cold spots. All of these measures and tools are descriptive and exploratory. To model spatial autocorrelation, more advanced techniques and models are needed.

REFERENCES

Anselin, L. (1995). Local Indicators of Spatial Association—LISA. *Geographical Analysis*, 27(2): 93–116.

Anselin, L. and Griffith, D. A. (1988). Do spatial effects really matter in regression analysis? *Papers of the Regional Science Association*, 65: 11–34.

Cliff, A. D. and Ord, J. K. (1973). *Spatial Autocorrelation*. London: Pion.

Cliff, A. D. and Ord, J. K. (1981). *Spatial Processes: Models and Applications*. London: Pion.

Getis, A. (1991). Spatial interaction and spatial autocorrelation: A cross-product approach. *Environment and Planning A*, 23: 1269–1277.

Getis, A. and Ord, J. K. (1992). The analysis of spatial association by use of distance statistics. *Geographical Analysis* 24(3): 189–207.

Griffith, D. A. (1996). Some guidelines for specifying the geographic weights matrix contained in spatial statistical models. In S. L. Arlinghaus and D. A. Griffith (eds.), *Practical Handbook of Spatial Statistics*. Boca Raton, FL: CRC Press, pp. 65–82.

Odland, J. (1988). *Spatial Autocorrelation*. Newbury Park, CA: Sage.

Tobler, W. R. (1970). A computer movie simulating urban growth in the Detroit region. *Economic Geography*, 46 (Supplement): 234–240.

INDEX